ENERGY
EFFICIENCY
in the URBAN
ENVIRONMENT

MECHANICAL and AEROSPACE ENGINEERING

Frank Kreith & Darrell W. Pepper
Series Editors

ENERGY EFFICIENCY in the URBAN ENVIRONMENT

Heba Allah Essam E. Khalil
Essam E. Khalil

CRC Press
Taylor & Francis Group
Boca Raton London New York

CRC Press is an imprint of the
Taylor & Francis Group, an **informa** business

CRC Press
Taylor & Francis Group
6000 Broken Sound Parkway NW, Suite 300
Boca Raton, FL 33487-2742

First issued in paperback 2019

© 2015 by Taylor & Francis Group, LLC
CRC Press is an imprint of Taylor & Francis Group, an Informa business

No claim to original U.S. Government works

ISBN-13: 978-1-4822-5063-3 (hbk)
ISBN-13: 978-0-367-37781-6 (pbk)

Visit the Taylor & Francis Web site at
http://www.taylorandfrancis.com

and the CRC Press Web site at
http://www.crcpress.com

To our students who deserve to know.

To our family who supported us to know.

To my mother who always urged me to know.

To my sons, may you enjoy a better environment.

Contents

List of Figures

List of Tables

Preface

Energy crisis, urbanisation and climate change are three global challenges of the twenty-first century. They are closely interrelated either by causal effect (where one causes the other) or by three separate phenomena that have parallel impacts.

This book is an attempt to study these challenges worldwide as well as to deal with or combat them in the Middle East – with a special focus on Egypt. In this quest, the authors address these issues from multiple perspectives, disciplines and scales. The first four chapters of this book address the macroscale of urbanism. First, the macroscale of cities is studied from the perspective of city dwellers' quality of life. Second, achieving energy efficiency through urban planning is investigated as a tool for improving city energy performance. Third, the energy efficiency performance of cities is studied by measuring various related indices and their indicators; this is tied to which sustainable urbanism principles these cities follow. Fourth, the author analyses how informal areas – as the most predominant feature of urbanisation in developing countries – achieve sustainable development as a different approach to sustainable urbanism. Case studies are presented and analysed in each chapter; these studies mainly use Egypt as an example for arid zones, developing countries and high rates of urbanisation and include Africa's biggest mega city: Cairo.

With its location and urban characteristics, Egypt represents a true model of the three challenges that the book addresses. It is continuously urbanising with its rural settlements expanding to become cities and its cities growing and encroaching on surrounding agricultural land. Egypt has scarce water resources because it is 95% arid. It is also highly susceptible to climate change, which has been clearly monitored in the past decade. Egypt's Nile Delta and north coast are among the most vulnerable risk zones worldwide. Finally, Egypt is witnessing a daily growing energy crisis with high consumption rates, no energy efficiency measures and continuously diminishing potential oil reserves. These factors make it crucial for Egypt to address these issues and to formulate and implement an energy efficiency policy to help mitigate the effects of urbanisation and climate change and to improve quality of life through energy-efficient urbanism.

The second part of the book addresses the challenges through the microscale of buildings and the perspective of ensuring indoor air quality within the boundaries of energy efficiency. Energy performance of buildings should include a general framework for the calculation of energy performance and building categories together with thermal characteristics of building, air-conditioning, ventilation, lighting and appliances aspects. These include the contribution of active solar systems to domestic water heating based on renewable energy sources, Combined Power and Heat (CPH) production and district cooling systems. The book demonstrates the importance of incorporating an energy performance directive as a standard in our region; such a goal will aid energy savings in large buildings and set regulations for energy-efficient designs that are based on standard calculation methods. Energy standards would be largely based on international standards and appropriately modified to suit local practices. The target is to develop standardised tools for the calculation of the energy performance of buildings, with defined system boundaries for the different building categories and for different cooling/heating systems, and to develop a common procedure for obtaining an 'energy performance certificate'. This book attempts to provide transparent information regarding output data (reference values, benchmarks, etc.) and to define comparable energy-related key values (kWh/m^2, kWh per person, kWh per apartment, kWh per produced unit etc.). Proposals to develop a common procedure for an energy performance certificate and CO_2 emissions are also given.

Acknowledgements

The authors thank their colleagues and collaborators for the useful discussions that led to the elaboration of this book. The authors express their special thanks and gratitude to their family legends, the late Dr. Khalil Hassan Khalil and late Dr. Galal Shawki of Cairo University, for their teaching and mentoring. The authors extend their gratitude to Dr. Ahmad Khalil for his extensive discussions and technical support. Thanks are also due to Eng. Hala Assaf for her everlasting support to Dr. Heba Allah Essam Khalil.

Heba Khalil extends her gratitude to her mentor, Dr. Sahar Attia, for her continuous encouragement in the publication of this book. Thanks are also due to her husband, Eng. Abdel Monem Abul Fadl; without his support, this book would not have been possible.

The authors also thank the publishers for their advice and support throughout the publishing process.

Finally, the authors thank one and all who inspired this work by posing questions or probing existing problems that required new research to realise new concepts and solutions.

About the Authors

Heba Allah Essam El-Din Khalil holds a BSc (2000) in architectural engineering, MSc (2003) in urban planning and a PhD (2007) in architectural engineering from Cairo University. Currently, she is an associate professor in the Department of Architectural Engineering, Faculty of Engineering, Cairo University, with 15 years of academic experience. She has pursued scientific research in various fields, including community development, participatory evaluation, informal areas development, energy-efficient strategies in urban planning, sustainable urbanism, green rating systems, affordable housing, quality of life and strategic planning. She has supervised more than 25 MSc and PhD theses in related research domains. She has participated in several national and international conferences, workshops and training courses. She is currently participating in a number of joint research projects with Cairo University and other international research institutes. She is a member of the informal hub of the Habitat Universities initiative UN-Habitat.

Dr. Khalil has additional professional experience as an architect and as an urban planner on various projects. She worked with international agencies as Deutsche Gesellschaft für Internationale Zusammenarbeit (GIZ) GmbH in participatory development, including facilitation and evaluation. She has worked for UN-Habitat in sustainable urban development strategy in Arab countries and in strategic planning for cities and city regions in Egypt. She has designed a number of private and public buildings. She was a member of the winning team of Khedive's Cairo rehabilitation competition. This multidisciplinary experience has supported her academic competence and sense of the real surrounding world.

Essam E. Khalil holds a BS (1971) and MS (1973) in mechanical engineering from Cairo University, and DIC (1976) and PhD (1977) from Imperial College of Science and Technology, London University, UK. Currently, he is a professor of mechanical engineering at Cairo University (since June 1988). He has more than 44 years of experience in design and simulation of combustion chambers for terrestrial and aerospace applications. He has published more than 600 papers in journals and conference proceedings as well as 12 books on combustion, energy and indoor air quality control. Dr. Khalil is a fellow of ASME, fellow of ASHRAE, convenor of ISO TC205WG2 on energy efficiency and ISO TC163 WG4, and also a member of ISO TC163. He is an ASHRAE distinguished lecturer. He is a recipient of the Outstanding Achievement in Science Award, Cairo University, 2005, 2006 and 2008; best Paper Award from AIAA Terrestrial Energy TC, 2005;

ASME George Westinghouse Award, 2009; ASME James Harry Potter Gold Medal 2012; and Distinguished Service Award ASHRAE and Sustained Services Award AIAA.

He is the editor of *Energy & Buildings, CFD Letters, and International Journal of Reacting Systems* and also an associate editor of *Energy Technology and Policy.*

1

Energy-Efficient Quality of Life

1.1 General

Energy-efficient quality of life will be viewed in this chapter within the following framework:

Energy efficiency concepts, definitions and measures

Quality of life as an indicator for human progress

Ecological Footprint and Human Development Index

Quality of life and sustainability

Energy performance

Energy efficiency indicators

Energy efficiency standards

Energy labels

The chapter ends with concluding remarks.

1.2 Energy Efficiency Concepts, Definitions and Measures

The year 2007 marked a watershed in human history; for the first time, more people lived in cities than in the countryside [1]. It is estimated that by the middle of the century, 70% of the world's population will reside in cities [2]. This rapid urbanisation is especially evident in the cities of the developing world where more than 70% of the world's urban population currently live.

Over the past 50 years, cities have expanded into the land around them at a rapid rate. Highways and transport systems have been built in tandem to support this physical growth. Valuable farmland has been eaten up, and car dependency has increased. Between 2010 and 2015, urban populations are expected to grow by around 200,000 people on average each day, with 91% of this daily increase expected to take place in developing countries. This increase is led by Asian and African cities [2].

UN-HABITAT's *State of the World's Cities* report for 2006–2007 points out that, in many cases, urban growth will become synonymous with slum formation. Already, Asia is home to more than half of the world's slum population (581 million) followed by sub-Saharan Africa (199 million) and Latin America and the Caribbean (134 million) [3]. Cities and urban settlements must be prepared to meet this challenge. To avoid being victims of their own success, cities must search for ways in which to develop sustainably.

A successful city must balance social, economic and environmental needs; it has to respond to pressure from all sides. It should offer investors security, infrastructure (including water and energy) and efficiency. It should also put the needs of its citizens at the forefront of all its planning activities. It must recognise its natural assets, its citizens and its environment and must build on these to ensure the best possible returns.

Modern cities are products of fossil fuel technology—most of the world's energy is used by cities themselves and by the farming, industrial production and transport systems that supply them. Modern urban living crucially depends on uninterrupted energy supplies. The world's major transport systems start and end in cities. They are the nodes from which mobility emanates, with low transport costs having rendered distances irrelevant, plugging cities into an increasingly global hinterland [4].

As world populations grow (many faster than the average 2%), the need for more and more energy is exacerbated as illustrated in Figure 1.1. The world's electricity consumption in 2011 was 22 TWh/yr and estimated to reach 23 TWh by 2020 [5]. Such ever-increasing demand could place significant strain on the current energy infrastructure and could potentially damage

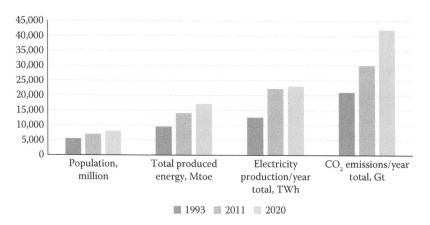

FIGURE 1.1
Key indicators for energy demand and CO_2 emissions. (From *World Energy Council, Energy for Tomorrow's World, the Realities, the Real Options and the Agenda for Achievement*, WEC Commission Report, WEC, New York, 1993; World Energy Council, *World Energy Resources: 2013 Survey*, WEC, London, 2013.)

world environmental health by CO, CO_2, SO_2, NO_x effluent gas emissions, and global warming. Achieving solutions to environmental problems that we face today requires long-term potential actions for sustainable development. Enhanced lifestyle and energy demand rise together where the wealthy industrialised economies, which contain 25% of the world's population, consume 75% of the world's energy supply [6]. The world's energy consumption in 2011 was 14,092 Mtoe; about 30 Giga metric tons of CO_2 emissions were released into the atmosphere to meet this energy demand [5]. Greenhouse gas (GHG) emissions and energy demand have risen high on the global environmental agenda—particularly with the Kyoto Protocol and other related global agreements. Consequently, an urgent need for the incorporation of energy efficiency issues to be included in urban planning and construction has risen [7]. Currently, it is the pursuit of more economic growth that guides development, especially in cities. However, there is a rising demand for an alternate view of development—one that provides a better quality of life and, at the same time, one that works within the constraints of limited available resources, especially energy. This chapter will review the current understanding of what comprises the notion of quality of life and its relation to sustainable development. Furthermore, this chapter investigates a number of the available indices used to measure quality of life and sustainable development. It reviews the different indicators for energy efficiency at the macrolevel and at the microlevel.

1.3 Quality of Life as an Indicator for Human Progress

Quality of life has been the domain of development discourse for the past decade. It has been widely recognised that measuring progress in terms of gross domestic product (GDP) is not sufficient [8–10]. This chapter attempts to examine the concept of quality of life with respect to other related concepts such as standard of living, cost of living and life satisfaction. In the quest to rate cities and countries according to their quality of life, numerous attempts by various organisations can be found. However, Cummins [11] states, 'The quality of life construct has a complex composition, so it is perhaps not surprising that there is neither agreed definition nor a standard form of measurement'.

Consequently, numerous debates exist regarding quantifying quality of life, such as those concerning what aspects should be measured, the relative weight of different aspects [12], the average versus real citizen's quality of life and objective versus subjective indicators. Khalil reviewed the efforts of different organisations to address the issue. She analysed the different aspects and relative weights used by these organisations, thus revealing multiple facets of the concept of quality of life [13].

1.3.1 Quality of Life versus Standard of Living Concepts

GDP (as a measure of the average standard of living) used to be considered the one and only indicator for evaluating the well-being of a nation. However, many concerns were raised with respect to inequality of wealth distribution; hence, the Gini coefficient was introduced to measure income distribution. Further, GDP was normalised using purchasing power parities (PPPs) to be able to compare results from various countries on an actual base with respect to the local circumstances of cost of living.

However, according to the Economist Intelligence Unit, it has long been accepted that material well-being, as measured by GDP per person, cannot alone explain the broader quality of life in a country [14]. Income or standard of living is of course crucial; without resources, any progress is difficult. Yet, other aspects must also be gauged. One strand of the literature has tried to adjust GDP by quantifying omitted facets—for example, various non-market activities and social ills such as environmental pollution. However, it has faced insurmountable difficulties in assigning monetary values to the various factors and intangibles of socioeconomic well-being.

Currently, Human Development Index (HDI) is another widely used indicator for quantifying progress. It combines measures of life expectancy, education and standard of living in an attempt to quantify the options available to individuals within a given society. It is used by the United Nations Development Programme in their Human Development Report.

In the 2010 Human Development Report, three new measures were introduced. The inequality-adjusted HDI, the Gender Inequality Index and the Multidimensional Poverty Index were applied to most countries in the world, and they provide important new insights [15].

Despite the high appreciation and dependence on HDI, there are still more aspects that can describe and be attributed to quality of life. The debate is still unresolved. According to Veenhoven [12], HDI is of little weight as a measure of overall well-being. He claims that HDI adds apples and oranges, where chances for a good life (wealth and education) are added to outcomes (life expectancy) and outer qualities (wealth and equality) are added to an inner one (education).

Alternatively, there have been numerous attempts to construct alternative, non-monetary indices of social and economic well-being. The following section attempts to study the concept of quality of life and the relevant efforts to measure it worldwide.

1.3.2 Quality of Life Indices

According to Rapley [16], quality of life has developed from being a social scientific index of the relative well-being of whole populations (a measure of the state of states) to being a measurable aspect of individual subjective experiences (an index of the state of persons). The ecological economist

Robert Costanza adds that although quality of life has long been an explicit or implicit policy goal, adequate definition and measurement have been elusive [17].

Regarding measuring quality of life, Gabriel and Rosenthal criticise current media and academic measures. They claim that media rankings of city quality are largely ad hoc. Despite considerable progress having been made academically in measuring urban quality of life [18–22], the most comprehensive measures of quality of life have been static in nature [19,20]. Thus, little data is available concerning changes in those measures over time [23].

Following the longstanding controversy in social indicators research, two approaches for measuring quality of life are identified as 'objective' or social indicators and subjective well-being (SWB) [25].* Although the objective approach focuses on measuring 'hard' facts, such as income or living accommodations in meter square, the subjective approach is concerned with 'soft' matters such as satisfaction with income and perceived adequacy of dwelling [12]. These two contrary conceptualisations are tracked by Noll; his Scandinavian View† centred on the notions of 'good society' and the objective indicators of quality of life (or living) of society as a whole. On the other hand, subjective indicators at the level of individuals are more evident in the American model that commands more consensuses in the Western world [16]. Concerning the term subjective well-being, Veenhoven prefers the term life satisfaction because life satisfaction refers to an overall evaluation of life rather than to current feelings or to specific psychosomatic symptoms [29].

Some of the indices that focused on subjective surveys and questionnaires to measure quality of life include the Global Person Generated Index (GPGI) that uses a mix of open-ended questions, scoring and point allocation to establish a particular person's satisfaction with the most important areas of his life [30–32]. Another example is the Wellbeing in Developing Countries Quality of Life (WeDQoL) [33], which obtains scores that not only reflect the general perspective of people in each country investigated but also the priorities of each person completing the measure, taking into account their geographical and social positions [34]. World Health Organization Quality of Life (WHOQOL) is another example that is used in health-related quality of life research. It has a questionnaire about self-perceived well-being during the last two weeks and addresses physical health, psychological health, social relationships and environmental conditions, in addition to perceived overall quality of life [35].

Practically, there are indicators that cannot be graded as either subjective or objective because there are different gradings in between. This is evident in the four indices proposed and used by different organisations (and reviewed in the following sections) to score and rate cities and countries

---

* A history of the social indicators and SWB movements in the social sciences can be found in Land [24].
† Demonstrated in the works of Drewnowski [26]; Erikson and Uusitalo [27]; and Erikson [28].

according to their quality of life/living. These organisations vary in nature and interests and thus in focus and methodology. Mercer, a private company, is the global leader for trusted human resources and related financial advice, products and services. Their aim is to enhance the financial and retirement security, health, and productivity and employment relationships of the global workforce. The EIU is the business information arm of The Economist Group, publisher of *The Economist*. For more than six decades, the unit has been delivering impartial business intelligence to clients, equipping decision makers with insight they can trust. Another attempt independently rates nations according to quality of life. The Organisation for Economic Cooperation and Development (OECD) has a different focus. Currently, 34 countries are building on its successor, the Organisation for European Economic Cooperation (OEEC).* Today, its members regularly work together to identify, discuss and analyse problems and to promote policies.

The objective approach can be detected in the measurement of quality of living performed by Mercer consultants, while the subjective approach can be detected in OECD's measurements of quality of life [13]. Another possibility for combining objective and subjective approaches to measuring overall well-being is Happy Life Years (HLY) suggested by Veenhoven, where the subjective indicator called 'life satisfaction' is combined with objective length of life and is expressed in the number of 'happy life years' [12], which depends on the fit between environmental conditions and personal capabilities.

1.4 Ecological Footprint and Human Development Index

In the development realm, there are a number of indices used to measure or to quantify progress. The ecological footprint (EF) developed by Wackernagel and Rees is considered one of the most important indices used to quantify the impact of development [36]. It translates various types of consumption into the common metric: total area of productive land and water ecosystems required to produce the resources that the population consumes and to assimilate the wastes that the population produces, wherever on Earth that land and water may be located [37].

The shortcoming of EF is its limited ability to provide positive policies that enable cities to solve their local and global problems. In this stream, Newman emphasises the importance of a sustainability assessment approach for cities that take environmental impact seriously, while simultaneously asserting the value of social and economic progress [38]. On the other hand as mentioned before, one of the more widely used indices is HDI.

---

* OEEC was established in 1947 to run the US-financed Marshall Plan for reconstruction of a continent ravaged by war.

Despite the high appreciation of and dependence on HDI, the debate concerning its relevance for measuring quality of life is still unresolved. Moreover, there is no straightforward pattern relating HDI to other dimensions of human development such as sustainability and empowerment. The lack of correlation can be seen in the large number of countries that have high HDI values but perform poorly on sustainability; about a quarter of the world's countries have a high HDI [15].

As an approach to remedy the limitations of each index, the Global Footprint Network presents their Human Development Initiative as the combination of two indices: the EF, with data about given current population and available land area, and the United Nations' HDI, which measures a country's average achievements in areas of health, knowledge and standard of living [39]. This combination enables measurement of the minimal conditions for sustainable human development, defined as a situation in which all humans can have fulfilling lives without degrading the planet. This is illustrated in Figure 1.2, where all countries lie outside the sustainable quadrant. All countries either provide high living conditions but exceed the available earth's biocapacity, or do not overburden available resources but do not provide adequate living conditions for their population. Careful investments in energy systems, transportation, health, education or urban infrastructure can move countries into a stronger position in the HDI-EF graph, providing higher human development and less of an ecological deficit. These investments need to be tested for how they will affect the three sub-indices of the HDI, as well as the country's resource dependence. If they generate gains in both arenas, they will advance human well-being that can last [40].

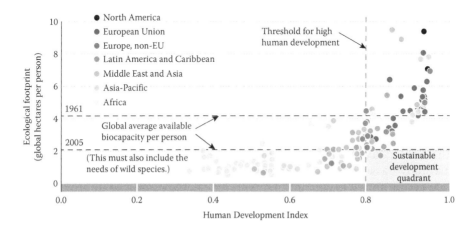

FIGURE 1.2
Human Development Index and ecological footprint, 2005. (From Our Human Development Initiative, 2011, Global Footprint Network, from http://www.footprintnetwork.org/en/index. php/GFN/page/fighting_poverty_our_human_development_initiative)

1.5 Quality of Life and Sustainability

Quality of life has been the domain of development discourse for the past decade. It has been widely recognised that measuring progress in terms of GDP is not sufficient. However, an adequate definition of quality of life is still missing. Diverse 'objective' and 'subjective' indicators across a range of disciplines and scales and recent work on SWB surveys and the psychology of happiness have spurred renewed interest [17].

Currently, there are a number of indices proposed and used by different organisations to score and rate cities and countries according to their quality of life. These organisations vary in nature and interests and thus in focus and methodology [13]. The most important indices are Quality of Living by Mercer [41], Quality of Life Index by the EIU [14],* nations' rankings according to quality of life compiled independently [42], and OECD's Your Better Life Index [43]. It is apparent that these different indices focus on a number of common aspects as the main core of quality of life, namely: housing, income, jobs, community, education, environment, governance, health, life satisfaction, safety and work–life balance.

Despite these various efforts, an important aspect is still missing in most indices. It is sustainability—whether or not this level of quality of life can be sustained and whether or not it is affecting the ability of future generations to attain such levels. This notion is still underscored because it is complex to assess [13].

There are other indices that measure and rate the sustainability of a city. One of the most recent sustainability indices is the Green City Index. This index measures the current environmental performance of major cities on different continents as well as their commitment to reducing their future environmental impact by way of on-going initiatives and objectives. The EIU, in cooperation with Siemens, developed the methodology. An independent panel of urban sustainability experts provided important insights and feedback on the methodology. The index scores cities across eight categories (as shown in Figure 1.3): CO_2 emissions, energy, buildings, transport, water, waste and land use, air quality and environmental governance, and 30 individual indicators. Sixteen of the index's 30 indicators are derived from quantitative data and aim to measure how a city is currently performing. The remaining 14 indicators are qualitative assessments of cities' aspirations or ambitions—for example, their commitment to consuming more renewable energy, to improving the energy efficiency of buildings, to reducing congestion or to recycling and reusing waste [44].

---

* The business information arm of The Economist Group, publisher of *The Economist*.

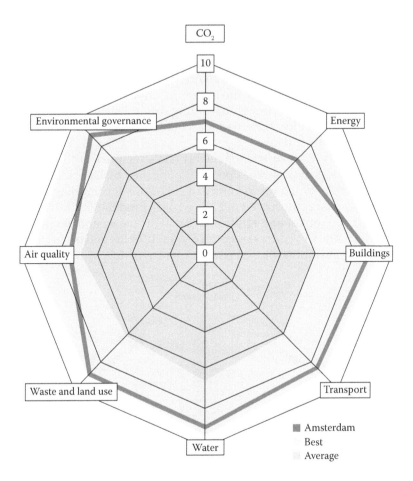

FIGURE 1.3
Green City Index for European cities. (From *Green City Index*, Siemens, Retrieved May 7, 2012, from http://www.siemens.com/entry/cc/en/greencityindex.htm.)

The index is composed of aggregate scores of all of the underlying indicators. It is first aggregated by category, creating a score for each area of infrastructure and policy (i.e. energy). Finally, overall, it is based on the composite of the underlying category scores. To create the category scores, each underlying indicator is aggregated according to an assigned weighting. In general, most indicators receive the same weighting, or importance, in the index. The category scores are then rebased onto a scale of 0–10. To build the overall index scores, the EIU assigned even weightings on each category score; the index is essentially the sum of all category scores, rebased out of 100. This equal weighting reflects feedback from the expert panel as well as wider research on measuring environmental sustainability, which indicated that all categories in this index merit equal weighting [44]. Currently, this index covers Europe, Latin America, Asia, the United States and Canada, Germany and Africa [45].

Thus, the index offers a tool to enhance the understanding and decision-making abilities of all those interested in environmental performance, from individual citizens through to leading urban policymakers. However, this index is not directly concerned with quality of life or with how people are satisfied with their lives.

The recent City Prosperity Index (CPI) has attempted to fill this gap in assessment indices. It was developed and published by UN-HABITAT in its report on the state of the world's cities in 2012–2013. It comprises of five different dimensions, as illustrated in Figure 1.4 [2]: productivity, infrastructure development, quality of life, equity and social inclusion, and environmental sustainability. The CPI addresses issues of quality of life and of sustainability. However, the indicators that constitute quality of life in the CPI are just

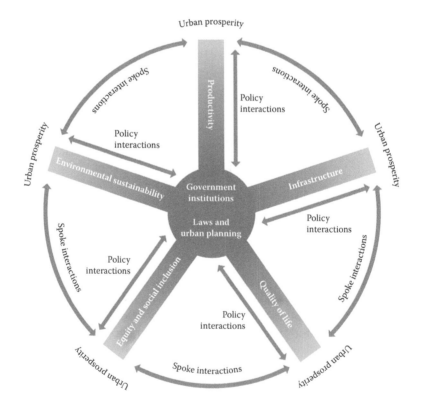

FIGURE 1.4
The Wheel of Urban Prosperity. (From United Nations Human Settlements Programme (UN-HABITAT), State of the World's Cities Report 2012/2013: Prosperity of Cities, UN-HABITAT, Nairobi, Kenya, 2012.)

a part of the overall indicators comprising the notion of quality of life as viewed by other indices.

This index is still under development by the secretariat of the CPI, an expert group formed by UN-HABITAT to discuss and develop the used indicators.* The main debates about indicators are which ones to measure, their relative weights, whether they represent the average versus real citizen's quality of life, and the use of objective versus subjective indicators. This is still not clearly resolved in the CPI, thus leaving much space for debating its representation of reality.

Another recent addition to the list is the International Ecocity Framework and Standards (IEFS) developed by the Ecocity Builders and the British Columbia Institute of Technology, Canada [46]. It is still under development but has similar categories and indicators as the Green City Index. However, it depends on more qualitative assessment rather than quantitative assessment, which can serve better when using participatory evaluation of environmental performance of districts and cities.

1.5.1 Quality of Life, Energy Efficiency and Renewable Energies in the Built Environment

Despite the involvement of the Green City Index in measuring energy efficiency as part of achieving environmental sustainability, it does not explicitly correlate to measuring how much energy is used to attain a certain quality of life for city dwellers. Thus, the indicators concerning energy efficiency in the Green City Index are first extracted from different categories and highlighted. As mentioned earlier, the Green City Index was originally subdivided into eight categories: CO_2 emissions, energy, buildings, transport, water, waste and land use, air quality and environmental governance, and 30 individual indicators. In each report, the EIU aggregated categories differently combining energy and CO_2, creating a separate category for land use or combining it with buildings, and creating a separate category for sanitation. A review of the indicators of each category in different reports and their relevance to energy efficiency resulted in the list of indicators shown in Table 1.1 (common indicators in different reports are highlighted) [44,47–50]. It is clear that similar indicators are used in all reports. However, other indicators are tailored to certain areas according to the state of things and to available information [51]. Second, these energy efficiency indicators can be classified into two groups: the first group relates directly to the energy used to achieve a good quality of life, while the other group relates indirectly to the matter and is more concerned with sustaining this quality of life from the point of using energy efficiently. Table 1.2 shows this classification of indicators.

---

* A series of Expert Group meetings started in May 2013 in Tehran to discuss the validity of the index, which continued in Nairobi in June 2013 for further development.

TABLE 1.1

Indicators Concerning Energy Efficiency Derived from the Green City Index for Different Areas

Category	Indicators	Description	Index
Energy and CO_2	Energy consumption	Total final energy consumption, in gigajoules per head	Europe; U.S. & Canada
	Electricity consumption per unit of GDP	Total final energy consumption, in megajoules per unit of real GDP (in Euros, base year 2000)	All
	Renewable energy consumption	The percentage of total energy derived from renewable sources as a share of the city's total energy consumption, in terajoules	Europe
	Clean and efficient energy policies	An assessment of the extensiveness of policies promoting the use of clean and efficient energy	All
	Climate change action plan	Measure of a city's strategy to combat its contribution to climate change	Latin America; Asia
	Access to electricity	Percentage of households with access to electricity	Africa
Buildings	Energy consumption of residential buildings	Total final energy consumption in the residential sector, per square meter of residential floor space	Europe
	Energy-efficient building standards	An assessment the extensiveness of cities' energy efficiency standards for buildings	Europe; U.S. & Canada
	Energy-efficient building initiatives	An assessment of the extensiveness of efforts to promote energy efficiency of buildings	Europe; U.S. & Canada
	Eco buildings policy	Measure of a city's efforts to minimise the environmental impact of buildings	Latin America; Asia
	Number of LEED-certified buildings	Number of LEED-certified buildings (silver, gold or platinum) per 100,000 persons	U.S. & Canada
Land use	Green land-use policies	An assessment of the comprehensiveness of policies to contain the urban sprawl and promote the availability of green spaces Assessment of a city's efforts to sustain and improve the quantity and quality (e.g. proximity and usability) of green spaces and its tree planting policy	Europe; U.S. & Canada
	Green spaces per capita	Sum of all public parks, recreation areas, greenways, waterways and other protected areas accessible to the public, in m² per inhabitant	All but Europe
	Population density	Population density, in persons per km²	All but Europe

	Land-use policy	Measure of a city's efforts to minimise the environmental and ecological impact of urban development	Latin America; Asia; Africa
	Urban sprawl	Assessment of how rigorously a city promotes containment of urban sprawl and reuse of brown field areas	U.S. & Canada
	Population living in informal settlements	Percentage of the population living in informal settlements	Africa
Transport	Use of non-car transport	The total percentage of the working population travelling to work on public transport, by bicycle and on foot	Europe; U.S. & Canada
	Size of non-car transport network	Length of cycling lanes and the public transport network, in km per m^2 of city area. Evaluation of availability of public transport, including length of public transport network	Europe; U.S. & Canada
	Green transport promotion	An assessment of the extensiveness of efforts to increase the use of cleaner transport	Europe; U.S. & Canada
	Length of mass transport network	Composed of two sub-indicators: (1) total length of all train, tram, subway, bus and other mass transport routes within the city's boundaries, measured in terms of the area of the city (in km/km^2); and (2) total length of all superior modes of public transport, for example, Bus Rapid Transport BRT System, trolleybus, tram, light rail and subway, measured in terms of the area of the city (in km/km^2)	Latin America; Asia; Africa
	Stock of cars and motorcycles	Total stock of cars and motorcycles, with half-weighting allocated to motorcycles, measured in terms of vehicles per person	Latin America
	Urban mass transport policy	Measure of a city's efforts to create a viable mass transport system as an alternative to private vehicles	Latin America; Asia; Africa
	Average commute time from residence to work	Average commute time from residence to work, in minutes	U.S. & Canada
	Congestion reduction policy	Measure of a city's efforts to reduce congestion	All

Source: Khalil, H. (2012). Sustainable Urbanism: Theories and Green Rating Systems, *10th Annual International Energy Conversion Engineering Conference, 48th AIAA/ASME/SAE/ASEE Joint Propulsion Conference & Exhibit*, Atlanta, GA.

TABLE 1.2

Energy Efficiency Indicators as Achieving Good Quality of Life and Sustaining It

	Energy Efficiency Indicators	
Category	Achieving Good Quality of Life	Sustaining Good Quality of Life
Energy and CO_2	Energy consumption	Same
	Electricity consumption per unit of GDP	Same
	Access to electricity	Renewable energy consumption
		Clean and efficient energy policies
		Climate change action plan
Buildings	Energy consumption of residential buildings	Energy-efficient building standards
		Energy-efficient building initiatives
		Eco building policy
		Number of LEED-certified buildings
Land use	Green land-use policies	Same
	Green spaces per capita	Land-use policy
	Population density	Urban sprawl
	Population living in informal settlements	
Waste transport	Size of non-car transport network	Waste recycling and reuse policy
	Length of mass transport network	Use of non-car transport
	Stock of cars and motorcycles	Green transport promotion
	Average commute time from residence to work	Urban mass transport policy
	Congestion reduction policy	

Note: LEED, Leadership in Energy and Environmental Design.

1.6 Energy Performance

Energy performance of buildings should include a general framework for the calculation of energy performance and building categories together with the thermal characteristics of the building, air conditioning, ventilation, lighting and appliances considered. These include active solar systems' contribution to domestic water heating based on renewable energy sources, Combined Power and Heat (CPH) production and district cooling systems. The following sub-section demonstrates the importance of incorporating an energy performance directive as a standard in the Middle East region; such a goal will aid in energy savings in large buildings and will set regulations for energy-efficient designs that are based on standard calculation methods. The proposed standard would be largely based on international standards and would be appropriately modified to suit local practices. The target is to develop standardised tools for the calculation of the energy performance of buildings, with defined system boundaries for the different building categories and for different cooling/heating systems.

The goal is to develop common procedures for obtaining an 'energy performance certificate'. The present work provides transparent information regarding output data (reference values, benchmarks etc.) and defines comparable energy-related key values (kWh/m², kWh per person, kWh per apartment, kWh per produced unit etc.). Proposals to develop a common procedure for an 'energy performance certificate' and CO_2 emissions are given here.

Analyses of energy performance would be initiated from a country's power generation pattern [52] (Figure 1.5), with energy consumption by sector and energy consumption by utility (as shown in Figures 1.6 and 1.7 for Egypt), as an example. In commercial buildings, air-conditioning systems can consume as much as 56% of the total energy used in the building. Therefore, it is a challenge to design an optimum HVAC airside system that provides comfort and air quality in air-conditioned spaces with efficient energy consumption. The air condition to be maintained is dictated by how the conditioned space is used and by the comfort of users. Therefore, the air conditioning embraces more than cooling or heating. Comfortable air conditioning is defined as 'the process of treating air to control simultaneously its temperature, humidity, cleanliness, and distribution to meet the comfort requirements of the occupants of the conditioned space' [53]. Air conditioning, therefore, includes the entire heat exchange operation in addition to the regulation of velocity, thermal radiation and quality of air, as well as the removal of foreign particles and vapours [53].

It is probable that one calculation method will not cover all aspects and building categories for the proposed directive. For some applications, more

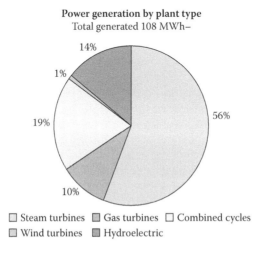

Power generation by plant type
Total generated 108 MWh–

14%
1%
19%
10%
56%

☐ Steam turbines ☐ Gas turbines ☐ Combined cycles
☐ Wind turbines ☐ Hydroelectric

FIGURE 1.5
Current power generation technologies in Egypt. (From Khalil, E. E. (2012). Energy performance of commercial buildings: A new direction, *Proceedings, ASME IMECE*, November.)

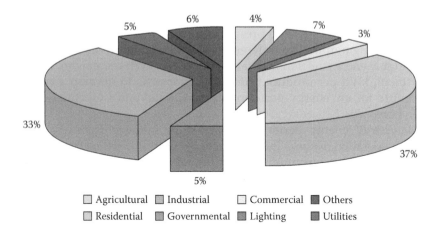

FIGURE 1.6
Energy consumption by sector 2006–2007 in Egypt. (From Khalil, E. E. (2012). Energy performance of commercial buildings: A new direction, *Proceedings, ASME IMECE*, November.)

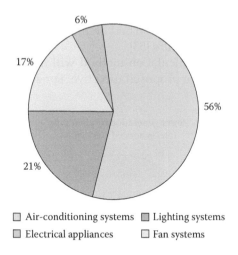

FIGURE 1.7
Building energy consumption by utility in Egypt. (From Khalil, E. E. (2012). Energy performance of commercial buildings: A new direction, *Proceedings, ASME IMECE*, November.)

advanced simulation models will have to be used to provide satisfactory
accuracy. The on-going and future work on methods for validation and on
documentation of simulation tools at Housing and Building Research Center
(HBRC) and Cairo University [52–54] could be valuable within the process of
approving models.

1.7 Energy Efficiency Indicators

Even now, the guidelines and design standards do not provide restricted utilisation strategies for conditioned air in building spaces. Indeed, this situation creates several inefficient systems and results in an expensive energy invoice. In some critical facilities, such as hospitals, HVAC designers face the problem of balancing the attainment of comfortable conditions with efficient energy utilisation. The relationship between HVAC system designs and optimum conditions and energy utilisation is still under investigation even today. Recent research [55,56] has investigated the effect of ventilation design on comfort and energy utilisation. The effect of the displacement ventilation on the humidity gradient in a factory located in a hot and humid region has been illustrated [55], showing the strong dependence between the correct supplying conditions and comfort. Indeed, displacement ventilation is recommended as an energy-efficient system; the resulting humidity and temperature gradient gives the designers the suitable tolerance for selecting more economical Air supply conditions [55]. In recent years, new ventilation system designs, such as under-floor systems, are becoming more popular for overcoming the problems of current systems. In office buildings, under-floor air supply is recommended as an alternative to ceiling air supply for overcoming the lack of flexibility in the ceiling systems and for improving the comfort conditions [55–59]. Actually, the energy utilisation mainly depends on the optimum utilisation of the conditioned air in the conditioned spaces. Currently, popular HVAC residential window and split units were further investigated in terms of energy costs and energy efficiency ratios (see Tables 1.3 through 1.8; where $1 = 7.05 LE).

As the world becomes increasingly dependent on electrical appliances and equipment, energy consumption rapidly rises every year. Many programmes have been established in various countries to increase end-use equipment energy efficiency. One of the most cost-effective and proven methods for increasing energy efficiency of electrical appliances and equipment is to establish energy efficiency standards and labels as defined in the following section.

TABLE 1.3

HVAC Unit Characteristics (Window Units)

Cooling Capacity, kW	Power Consumption, W	EER	COP
2.64	1023	8.8	2.58
3.66	1429	8.75	2.56
5.28	2000	9.0	2.64
7.03	2697	8.9	2.61

Note: EER = energy efficiency ratio; COP = coefficient of performance.

TABLE 1.4

HVAC Unit Costs (Window Units)

Cooling Capacity, kW	Power Consumption, W	Air Flow[a]	Unit Cost, LE
2.64	1023	540	1400
3.66	1429	630	1700
5.28	2000	1020	2000
7.03	2697	1275	2400

[a] Airflow is in m^3/h.

TABLE 1.5

HVAC Unit Estimated Energy Costs (Window Units)

Cooling Capacity, kW	Power Consumption, kWh/day at 40°C Outdoor	Air Flow	HVAC Cost, LE/day
2.64	10.23	540	2.046
3.66	14.29	630	2.858
5.28	20.00	1020	4.0
7.03	26.97	1275	4.394

Note: Calculated at 10 h/day and 0.2 LE/kWh.

TABLE 1.6

HVAC Unit Characteristics (Split Units)

Cooling Capacity, kW	Power Consumption, W	EER	COP
2.78	1056	9.0	2.64
3.66	1389	9.0	2.64
5.57	2111	9.0	2.64
7.03	2775	8.65	2.64
8.8	3409	8.8	2.58
10.26	3784	9.25	2.71

Note: EER = energy efficiency ratio; COP = coefficient of performance.

TABLE 1.7

HVAC Unit Costs (Split Units)

Cooling Capacity, kW	Power Consumption, W	Air Flow[a]	Unit Cost, LE
2.64	1056	509	3400
3.66	1389	509	3700
5.28	2111	798	4000
7.03	2775	956	4400
8.8	3409	1528	5500
10.26	3784	1528	6000

[a] Airflow is in m^3/h.

TABLE 1.8

HVAC Unit Estimated Energy Costs (Split Units)

Cooling Capacity, kW	Power Consumption, kWh/day at 40°C Outdoor	Air Flow m³/h	HVAC Cost, LE/day
2.64	10.56	509	2.112
3.66	13.89	509	27.78
5.28	21.11	798	4.222
7.03	27.75	956	5.550
8.8	34.09	1528	6.818
10.26	37.84	1528	7.568

Note: Calculated at 10 h/day and 0.2 LE/kWh.

1.8 Energy Efficiency Standards

Energy efficiency standards are a set of procedures and regulations that prescribe the energy performance of manufactured products, sometimes prohibiting the sale of products less energy efficient than the minimum standard. The term 'standard' commonly encompasses two possible meanings:

1. A well-defined protocol (or laboratory test procedure) for obtaining a sufficiently accurate estimate of the energy performance of a product in the way it is typically used or at least a relative ranking of the energy performance compared to other models
2. A target limit on energy performance (usually a maximum use or minimum efficiency) formally established by a government-based agency upon a specified test standard

1.9 Energy Labels

Energy efficiency labels are affixed to manufactured products indicating a product's energy performance (usually in the form of energy use, efficiency and/or cost) in order to provide consumers with the data necessary for making informed purchases. Energy labels serve as a complement to energy standards. These labels provide consumers with information that allows them to select more efficient models. Labels also allow utility companies and government energy conservation agencies to offer incentives to consumers to buy the most energy-efficient products. The effectiveness of energy labels is highly dependent on how the information is presented to the consumer. In developing countries, energy efficiency plays an important role in achieving global sustainable development. Energy efficiency improvements can slow the growth of energy consumption, can save consumers money and can reduce capital expenses for energy infrastructure. Energy consumption is increasing rapidly in these countries, yet energy efficiency remains far below the levels in developed countries. For most developing countries, the foreign exchange needed to finance energy sector expansion is a significant drain on reserves. Additionally, energy efficiency reduces local environmental impacts, such as water and air pollution from power plants, and mitigates greenhouse gas emissions. Standards and labelling programmes provide enormous energy-saving potential that can direct developing countries towards sustainable energy use. Improved end-use efficiency from standards and labelling programmes can contribute significantly to developing economies. The main benefits are

1. Less of a need for new power plants: The cost of saving 1 kWh of energy through energy efficiency programmes has proven much less expensive than producing 1 kWh of energy by building a new power plant.

2. Reduced greenhouse gas emissions: Less energy production means less carbon dioxide emissions from power plants. This contributes to environmental benefits such as slowing down environmental pollution and global warming and preserving natural resources and the ecosystem.

3. Improved competitiveness for local manufacturers: Local companies that upgrade the efficiency of their products can compete better with multinational companies, especially with lower production costs.

4. Higher consumer disposable income: Less spending on electric bills increases consumer purchasing power for other products, which helps local businesses.

5. Increased cash flow in the local economy: With higher disposable income, consumers are more willing to spend, thus injecting more money into the local economy.

6. Improved trade balance: Decrease in energy demand will reduce the consumption of indigenous resources (i.e., natural gas and oil), allowing more to be exported (for Lebanon, less to be imported). Increased export earnings (or less import spending) help alleviate trade deficits in Arab countries.

7. Avoiding a future energy deficit as power demand rises: Energy exporting countries have become net importers due to dramatic increases in electricity demand. Energy efficiency programmes can help slow down the demand and prevent energy deficits in the future.

Table 1.9 summarises recent international energy labelling activities that have impacted national economies.

TABLE 1.9

Summary of Pioneering International Programmes and Their Achievements

Country or Region	Programme	Achievements
Australia	Mandatory Standards and Labelling	11% reduction in energy consumption of labelled appliances in 1992; equals approximately 94 GWh of saved energy or a 1.6% decrease in total household electricity consumption
Europe	Mandatory Standards and Labelling	Germany: 16.1% increase in market efficiency (1993–1996);
		Netherlands: 12.6% increase in market efficiency (1992–1995);
		United Kingdom: 7.3% increase in refrigerator/freezer efficiency (1994–1996)
Philippines	Mandatory Standards and Labelling	25% increase in average efficiency of all air conditioners (after first year); energy savings: 6 MW in demand and 17GWh in consumption (after first year)
Egypt	Mandatory Standards and Labelling	10% decrease in refrigerator energy consumption (after 3 years);
		20% decrease in air conditioner energy consumption (after 3 years)
Thailand	Voluntary Labelling	14% decrease in refrigerator energy consumption (after 3 years);
		energy savings: 65 MW in demand and 643 GWh in consumption
United States	Mandatory Standards and Labelling	98% increase in refrigerator efficiency (1972–1988); more than 3% reduction in U.S. annual residential consumption from appliances and lighting equipment

A national standards and labelling programme is defined as a set of elements that ensure that energy efficiency standards and labelling efforts are effective, appropriate, strengthened over time and sustained. The building blocks fall into two categories, technical/policy and process. They include

1. Technical/Policy

   a. Accredited testing facilities: Facilities should be internationally accredited, should be staffed with competent testing personnel and have the capacity to test models in a timely manner.

   b. Appropriate testing procedures: Testing procedures are the methods by which the energy efficiency level of a product is deduced. The selected procedures should reasonably reflect a country's usage patterns and climate. This builds consumer confidence that test results accurately reflect the energy usage one will experience.

   c. Energy labels: Standards and labels can be established separately or as complementary programmes. Many types of labelling programmes exist.

   d. Energy efficiency standards can be mandatory or voluntary and can be based either on maximum energy consumption or on minimum energy efficiency.

   e. An energy policy framework that is conducive to energy efficiency is critical to the longevity of a national standards programme. Supportive policies include government procurement requirements, voluntary programmes, incentives to manufacturers, consumer awareness campaigns and demand-side management and integrated resource planning.

2. Proposed Process

   a. Compliance with voluntary and mandatory standards and labelling requirements must be ensured through a credible enforcement scheme to guarantee programme effectiveness. Programme evaluation will inform necessary programme modifications, justify further activities and provide the documentation necessary to sustain the standards and labelling programmes over the long term.

   b. The legislative process should ensure that standards and labels are periodically reviewed and raised ('ratcheted' upward) as the overall product efficiency of the market improves. The changes will mostly depend on the results of programme evaluation.

   c. In the programme design and improvement process, input from all stakeholders (government, private companies, consumer associations etc.) should be considered. Cooperation among the stakeholders is the key to programmes' success. However, local and national governments must also hold its decision final, after carefully considering all suggestions.

1.10 Concluding Remarks

It is important to incorporate an energy performance directive as a standard in this region. Meeting such a goal will aid energy savings in large buildings and will set regulations for energy-efficient designs that are based on standard calculation methods. The following recommendations apply:

1. Develop standardised tools for the calculation of the energy performance of buildings within urban and rural environments and their relation to providing adequate quality of life

2. Define system boundaries for the different building categories and different heating/cooling systems

3. Prepare requirement models for indoor air quality, thermal comfort in winter and summer, visual comfort and so forth

4. Define comparable energy-related key values (kWh/m², kWh per person, kWh per apartment, kWh per produced unit, etc.) and develop a common procedure for an 'energy performance certificate'

5. Design, construct and operate a solar decathlon (building) that can meet the rural and desert requirements and that can save diminishing fossil fuel resources

References

1. GlobeScan and MRC McLean Hazel. (2007). *Megacity Challenges: A Stakeholder Perspective*, Siemens AG, Munich.
2. United Nations Human Settlements Programme (UN-HABITAT). (2012). *The State of the World's Cities Report 2012/2013: Prosperity of Cities*, UN-HABITAT, Nairobi, Kenya.
3. United Nations Human Settlements Programme (UN-HABITAT). (2006). *The State of The World's Cities Report 2006/2007: 30 Years of Shaping The Habitat Agenda*, Earthscan, London, UK.
4. Girardet, H. (2004). *Urban Planning and Sustainable Energy: Theory and Practice*, Forum Barcelona, Retrieved May 12, 2009, from http://www.barcelona2004.org/www.barcelona2004.org/eng/banco_del_conocimiento/dialogos/fichac390.html
5. World Energy Council. (2013). *World Energy Resources: 2013 Survey*, World Energy Council, London.
6. Omer, A. M. (2008). Energy, environment and sustainable development, *Renewable and Sustainable Energy Reviews*, Vol. 12, No. 9, pp. 2265–2300.
7. United Nations Environment Programme. *Activities: Urban—Energy for Cities*, Retrieved May 14, 2009, from http://www.unep.or.jp/Ietc/Activities/Urban/energy_city.asp
8. Eckersley, R. (1998). Perspectives on progress: Economic growth, quality of life and ecological sustainability, In R. Eckersley, ed., *Measuring Progress: Is Life Getting Better?* CSIRO Publishing, Melbourne, pp. 3–34.
9. Shea, W. (1976). Introduction: The quest for a high quality of life, In J. King-Farlow and W. Shea, eds., *Values and The Quality of Life*, Science History Publications, New York, pp. 1–5.
10. Cummins, R. A., Eckersley, R., Pallant, J., Vugt, J. V., and Misajon, R. (2003). Developing a national index of subjective wellbeing: The Australian unity wellbeing index, *Social Indicators Research*, Vol. 64, pp. 159–190.
11. Cummins, R. (1997). *Comprehensive Quality of Life Scale—Adult*, School of Psychology, Deakin University, Melbourne.

12. Veenhoven, R. (2007). Subjective measures of well-being, In M. McGillivray, ed., *Human Well-Being, Concept and Measurement*, Palgrave/McMillan, Houndmills, NH, pp. 214–239.
13. Khalil, H. (2012). Enhancing quality of life through strategic urban planning, *Sustainable Cities and Society*, Vol. 5, pp. 77–86.
14. The Economist Intelligence Unit (EIU). (2007). The Economist Intelligence Unit's quality-of-life index, the world in 2005, *The Economist*, Retrieved January 20, 2012, from http://www.economist.com/media/pdf/QUALITY_OF_LIFE.pdf
15. United Nations Development Programme (UNDP). (2010). *Human Development Report 2010, 20th Anniversary Edition, The Real Wealth of Nations: Pathways to Human Development*, UNDP, New York.
16. Rapley, M. (2003). *Quality of Life Research: A Critical Introduction*, Sage, London.
17. Costanza, R., Fisher, B., Ali, S., Beer, C., Bond, L., Boumans, R., Danigelis, N. L., et al. (2008). An integrative approach to quality of life measurement, research and policy, *Surveys and Perspectives Integrating Environment & Society*, Vol. 1, pp. 11–15.
18. Roback, J. (1982). Wages, rents, and the quality of life, *Journal of Political Economy*, Vol. 90, pp. 1257–1278.
19. Blomquist, G., Berger, M., and Hoehn, J. (1988). New estimates of the quality of life in urban areas, *American Economic Review*, Vol. 78, No. 1, pp. 89–107.
20. Gyourko, J. and Tracy, J. (1991). The structure of local public finance and the quality of life, *Journal of Political Economy*, Vol. 99, No. 4, pp. 774–806.
21. Kahn, M. (1995). A revealed preference approach to ranking city quality of life, *Journal of Urban Economics*, Vol. 38, pp. 221–235.
22. Gabriel, S. A., Mattey, J., and Wascher, W. L. (2003). Compensating differentials and evolution in the quality-of-life among U.S. states, *Regional Science and Urban Economics*, Vol. 33, No. 5, pp. 619–649.
23. Gabriel, S. A. and Rosenthal, S. S. (2004). Quality of the business environment versus quality of life: Do firms and households like the same cities? *The Review of Economics and Statistics*, Vol. 86, No. 1, pp. 438–444.
24. Land, K. C. (1996). Social indicators and the quality of life: Where do we stand in the mid-1990s? *Social Indicators Network News*, Vol. 45, pp. 5–8.
25. Diener, E. and Suh, E. (1997). Measuring quality of life: Economic, social, and subjective indicators, *Social Indicators Research*, Vol. 40, pp. 189–216.
26. Drewnowski, J. (1974). *On Measuring and Planning the Quality of Live*, Mouton, published for the Institute of Social Studies, The Hague.
27. Erikson, R. and Uusitalo, H. (1987). The Scandinavian approach to welfare research, In R. Erikson, E. Hansen, S. Ringen, and H. Uusitalo, eds., *The Scandinavian Model: Welfare States and Welfare Research*, M. E. Sharpe, New York, pp. 177–193.
28. Erikson, R. (1993). Descriptions of inequality: The Swedish approach to welfare research, In M. Nussbaum and A. Sen, eds., *The Quality of Life*, Clarendon Press, Oxford, pp. 67–87.
29. Veenhoven, R. (1996). The study of life satisfaction, In W. Saris, R. Veenhoven, and A. Scherpenzeel, eds., *A Comparative Study of Satisfaction with Life in Europe*, Eötvös University Press, Budapest, pp. 11–48.
30. Ruta, D. (1998). Patient generated assessment: The next generation, *MAPI Quality of Life Newsletter*, Vol. 20, pp. 461–489.

31. Ruta, D. A., Garratt, A. M., Leng, M., Russell, I. T., and MacDonald, L. M. (1994). A new approach to the measurement of quality of life: The patient generated index (PGI), *Medical Care*, Vol. 32, No. 11, pp. 1109–1126.

32. Ruta, D., Camfield, L., and Martin, F. (2004). Assessing individual quality of life in developing countries: Piloting a global PGI in Ethiopia and Bangladesh, *Quality of Life Research*, Vol. 13, No. 9, p. 1545.

33. Woodcock, A., Camfield, L., McGregor, J. A., and Martin, F. (2009). Validation of the WeDQoL-goals-Thailand measure: Culture-specific individualized quality of life, *Social Indicators Research*, Vol. 94, No. 1, pp. 135–171.

34. McGregor, J. A., Camfield, L., and Woodcock, A. (2009). Needs, wants and goals: Wellbeing, quality of life and public policy, *Applied Research Quality Life*, Vol. 4, pp. 135–154.

35. WHOQOL Group. (1998). Development of the World Health Organization WHOQOL-BREF quality of life assessment, *Psychological Medicine*, Vol. 28, pp. 551–558.

36. Wackernagel, M. and Rees, W. (1996). *Our Ecological Footprint: Reducing Human Impact on the Earth*, New Society Publishers, Gabriola Island, BC.

37. Rees, W. (2000). Eco-footprint analysis: Merits and brickbats, *Ecological Economics*, Vol. 32, No. 3, pp. 371–374.

38. Newman, P. (2006). The environmental impact of cities, *Environment and Urbanization*, Vol. 18, No. 2, pp. 275–295.

39. *Our Human Development Initiative*. (2011). Global Footprint Network, Retrieved April 30, 2012, from http://www.footprintnetwork.org/en/index.php/GFN/page/fighting_poverty_our_human_development_initiative/

40. Wackernagel, M. (2009). Securing human development in a resource-constrained world, *DAC News*, Retrieved April 30, 2012, from http://www.oecd.org/dataoecd/52/1/43844294.htm#H56

41. Mercer. (2011). *Mercer 2011 Quality of Living Survey Highlights—Defining 'Quality of Living'*, Mercer, Retrieved January 25, 2012, from http://www.mercer.com/articles/quality-of-living-definition-1436405

42. Nation Ranking. (2011). *Quality of Life Index 2011 Ranking: Quantifying the World of Sovereign States*, Retrieved January 21, 2012, from https://nationranking.wordpress.com/category/quality-of-life-index/

43. Organization for Economic Cooperation and Development (OECD). (2011). *Better Life Initiative Executive Summary*, OECD Better Life Initiative, Retrieved January 20, 2012, from http://oecdbetterlifeindex.org/

44. The Economist Intelligence Unit. (2009). *European Green City Index: Assessing The Environmental Impact of Europe's Major Cities*, Siemens AG, Munich.

45. *Green City Index*. (2012). Siemens, Retrieved May 7, 2012, from http://www.siemens.com/entry/cc/en/greencityindex.htm

46. Ecocity Standards. (2011). *International Ecocity Framework and Standards*, Retrieved January 15, 2014, from http://www.ecocitystandards.org/

47. The Economist Intelligence Unit. (2010). *Latin American Green City Index: Assessing The Environmental Performance of Latin America's Major Cities*, Siemens AG, Munich.

48. The Economist Intelligence Unit. (2011). *African Green City Index: Assessing The Environmental Performance of Africa's Major Cities*, Siemens AG, Munich.

49. The Economist Intelligence Unit. (2011). *Asian Green City Index: Assessing The Environmental Performance of Asia's Major Cities*, Siemens AG, Munich.

50. The Economist Intelligence Unit. (2011). *US and Canada Green City Index: Assessing The Environmental Performance of 27 Major US and Canadian Cities*, Siemens AG, Munich.
51. Khalil, H. (2012). Sustainable urbanism: Theories and green rating systems, *10th Annual International Energy Conversion Engineering Conference, 48th AIAA/ASME/SAE/ASEE Joint Propulsion Conference & Exhibit*, Atlanta, GA.
52. Khalil, E. E. (2005). Energy performance of buildings directive in Egypt: A new direction, *HBRC Journal*, Vol. 1, pp. 197–213.
53. Medhat, A. A. and Khalil, E. E. (2006). Thermal comfort meets human acclimatization in Egypt, *Proceeding of Healthy Building*, Vol. 2, pp. 25.
54. Khalil, E. E. (2012). Energy performance of commercial buildings: A new direction, *Proceedings, ASME IMECE*, November 2012.
55. Kosonen, R. (2002). Displacement ventilation for room air moisture control in hot and humid climate, *ROOMVENT 2002*, pp. 241–244.
56. Leite, B. C. C. and Tribess, A. (2002). Analysis of under floor air distribution system: Thermal comfort and energy consumption, *ROOMVENT 2002*, pp. 245–248.
57. Kameel, R. and Khalil, E. E. (2002). Prediction of turbulence behavior using k-ε model in operating theatres, *ROOMVENT 2002*, pp. 73–76.
58. Khalil, E. E. (2008). Arab-air conditioning and refrigeration code for energy-efficient buildings, *Arab Construction World*, Vol. 28, No. 8, pp. 24–26.
59. Khalil, E. E. (2008). Air conditioning and refrigeration code for energy-efficient buildings in the Arab world, *Journal of Kuwait Society of Engineers*, Vol. 100, pp. 94–95.

2

Energy Efficiency Strategies in Urban Planning of Cities*

2.1 Introduction

As discussed in Chapter 1, the world's energy consumption in 2011 was 14,092 Mtoe; about 30 Giga metric tons of CO_2 emissions were released in the atmosphere to meet this energy demand [2]. Greenhouse gas (GHG) emissions and energy demand have risen high on the global environmental agenda—particularly with the Kyoto Protocol and other related global agreements. Consequently, an urgent need has arisen for the incorporation of energy efficiency issues into urban planning and construction [3]. To meet the urban challenges of today, and the challenges to come, appropriate planning strategies and management frameworks must be available, through which cities can apply innovative approaches suitable for their local circumstances. This chapter will review the challenges that cities face and factors that affect new strategies for urban planning where energy efficiency is the core issue shaping the city's future.

2.2 Cities and Energy Consumption: The Macrolevel

The city can be seen as an ecosystem comprising five main sub-systems that interact together. These are population sector, employment sector, housing sector, transport sector and urban land sector [4].

2.2.1 Size

Cities vary in size, starting from only 25,000 inhabitants—the number of city dwellers specified by Egypt's General Organization of Physical Planning (GOPP). In Egypt, population size is the main driving force of urbanisation

---

* Most of this chapter is derived from Khalil [1].

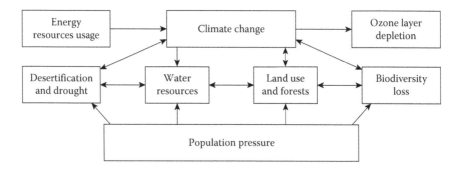

FIGURE 2.1
Relationships among desertification, climate change and biodiversity. (Adapted from The United Nations Economic Commission for Africa (UNECA) and North Africa Office, The fight against desertification and drought in North Africa, *The Eighteenth Meeting of the Intergovernmental Committee of Experts,* United Nations Economic Commission for Africa, Tangiers, Morocco, 2008.)

and in the quest for fulfilling their needs in an urban context. Another dimension related to population pressure in urban contexts is a city's growing size and its impact on climatic change. This is only one part of the environmental chain of energy use, ozone depletion, desertification and biodiversity loss, as shown in Figure 2.1 [5].

2.2.1.1 Mega Growth, Mega Complexity

The megacity is a relatively new form of urban development. In 1950, there were only two cities with populations of more than 10 million: New York and Tokyo. By 1975, two more locations, Shanghai and Mexico City, joined the club. However, by 2004, the number of megacities had rocketed to 22 and, together, these cities now account for 9% of the world's urban population. It is important to note that:

1. Mega cities' importance in the national and global economy is disproportionately high.
2. City governance has to adapt to the challenge of delivering holistic solutions across vast metropolitan regions.
3. City managers must strike the balance between three overriding concerns: economic competitiveness, environment and quality of life for urban residents.

Urban growth is spread unequally around the world, and the same is true of its largest cities. Most of the megacities in the developed world are growing slowly, if at all. Tokyo remains the largest with 35 million inhabitants, but the fastest growth will be in the developing world (particularly in Asia and Africa), placing huge pressure on infrastructure in those locations. By 2020, Mumbai, Delhi, Mexico City, São Paulo, Dhaka, Jakarta and Lagos will each have populations of more than 20 million. Moreover, it is estimated that between 2010 and 2015 some 200,000 people on average will be added to the world's urban population every day with 91% of this increase expected to take place in developing countries [6]. For many emerging cities, soaring populations are extremely difficult to manage; at current rates of growth, the number of inhabitants in Nigeria's Lagos will double by 2020, mainly through expansion of informal settlements. By contrast, most mature cities (as well as many transitional ones) will need to address a different kind of demographic challenge in the form of population ageing.

There is a continuous debate about megacities. On one level, these super-sized cities are seen as the engines of the global economy, efficiently connecting the flow of goods, people, culture and knowledge. They offer, at least potentially, unprecedented concentrations of skills and technical resources that can bring increased wealth and improved quality of life to vast numbers of people. However, megacities also conjure up an altogether darker vision. Most cities in the developing world face huge challenges ranging from congestion and pollution to security threats and inadequate services groaning under the weight of excessive demand. Those in the developing world also struggle to cope with the rapid growth of informal settlements. In 2006, almost one in three members of the world's urban population lives in slums, without access to good housing or basic services [7].

Today's megacities are not only bigger than the cities of the mid-20th century but also more complex. For one, they are increasingly competing with, and dependent on, relationships with other cities in the global economy. At the same time, we are witnessing the emergence of new city regions—sprawling conurbations that extend far beyond the boundaries of a single city. Examples include the 'BosWash stretch' (extending from Boston, MA, to Washington, DC) in the United States, and Chongqing in China. These huge megacity regions create a new urban dynamic. Commuters travel large distances from densely populated suburbs. Economic activity frequently becomes de-concentrated, dissipating from the centre to the periphery. Often fragmented systems of metropolitan governance have not caught up with this trend, with the result that it is difficult to deliver an efficient, holistic approach to infrastructure challenges at a metro regional level [8]. In addition, other new spatial configurations are increasingly taking place, such as urban corridors and city regions. These large urban configurations, as grouped in networks of cities, amplify the benefits of economies of agglomeration, increasing efficiencies and enhancing connectivity. They also generate economies of scale that are beneficial in terms of labour markets, as well as transport and communication infrastructure, which in turn increase local consumer demand [6].

2.2.2 Role and Competitiveness

In the context of continuous globalisation, there is a focus on competitiveness to attract investments to increase cities' prosperity. In this quest, there is a struggle among economic competitiveness and employment, environment and quality of life.

Megacities prioritise economic competitiveness and employment. In a study of which issues drive decision making in 25 megacities around the world, 81% of stakeholders involved in city management cited the importance of the economy and employment. There is a strong focus on creating jobs, with unemployment emerging as the top economic challenge for survey respondents from emerging and transitional cities. Competitiveness in the global economy is another important consideration. Six in ten stakeholders think that their cities place a high importance on making themselves competitive to attract private investment when deciding infrastructure issues [8].

Despite this inclination towards economic competitiveness, development decisions often involve difficult trade-offs between growth and greenness or growth and quality of life. There are obvious interdependencies among the three concerns. Competitive cities are more likely to have the wealth and resources to invest in high-quality infrastructure and services and to create economic and social opportunities for large numbers of the urban population. All things being equal, environmentally clean, modern cities are more attractive locations for a broad spectrum of business activities than those with heavy pollution. Equally, cities with a healthy, well-educated urban population are better positioned to attract investment than those where deprivation and inequality block a large portion of the population from participating in economic growth. This suggests that, in the long run, focusing on one of these concerns to the detriment of the others will be a recipe for failure (as shown in Figure 2.2) [8].

Therefore, cities need modern, efficient infrastructures, especially transportation networks. Abundant (and preferably skilled) labour together with modern information and communications technologies are also hugely important, as evidenced by the offshoring trend that has itself fuelled the growth of cities such as Bangalore in India. Another crucial factor is the

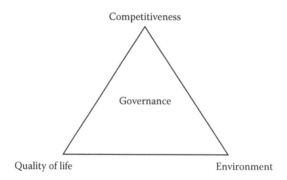

FIGURE 2.2
Striking a balance among quality of life, competitiveness and environment should be the main concern of megacities' governance.

quality of basic services; people with access to quality housing, education and good basic services such as water and electricity are much more likely to fulfil their potential and to contribute to economic growth. The wider business environment is also a key factor; research from the *Economist* intelligence unit indicates that clear, business-friendly policies and regulations are more important factors in attracting international investment than incentives such as subsidies and tax breaks [8].

Whatever their potential, however, many of today's megacities feature a catalogue of environmental problems. Congestion, air and water pollution, waste management and degradation of green areas are familiar issues in most large cities around the world, and they are particularly extreme in the megacities of the developing world. There are also huge inequalities in the distribution of wealth and in economic opportunity among cities. In its recent report on urbanisation trends, UN-Habitat describes cities as 'the new locus of poverty'. World Bank estimates predict that although rural areas are currently home to a majority of the world's poor, by 2035 cities will become the predominant locations of poverty [8].

The consequences of a failure to improve quality of life for the urban poor are huge. The UN-Habitat research indicates that people living in slums, where a large proportion of the urban poor reside, are more likely to be affected by child mortality and acute respiratory illnesses and by water-borne diseases than are their non-slum counterparts. They are also more likely to live near hazardous locations, making them more vulnerable to natural disasters such as floods. Inadequate access to basic services saddles them 'with heavy health and social burdens, which ultimately affect their productivity' [7]. Poverty may be less extreme in the more developed cities, but social problems still abound.

Historically, cities tend to get rich first and then clean up later. Unfortunately, that approach could be disastrous in the context of climate change; this is one reason for the growing focus on sustainable urban development. Sustainable solutions promote greater use of alternative energy sources and more energy-efficient buildings and transport, measures to combat congestion and CO_2 emissions, water and waste recycling, and the use of vegetation to filter pollution and capture carbon dioxide. Although several cities have started implementing at least some of these measures to good effect, there will be a need for more concerted efforts if the environmental cost of urbanisation is to be reduced.

Consequently, it is not growth and economic prosperity that cities should seek; it is rather a more sustainable development that combines efficiency, accountability and environmental responsiveness. This is a goal that comprises the main core of the Sustainable Cities Programme (SCP), a worldwide technical cooperation activity of the United Nations. The SCP works at the city level in collaboration with local partners to strengthen their capabilities for environmental planning and management. Each city-level SCP project is adapted to the particular needs, priorities and circumstances of that city;

nonetheless, all SCP city projects follow the same general approach, and all are implemented through the same series of activities known as the SCP process [9].

The SCP recognises that environmental deterioration is not inevitable. Although many, perhaps even most, cities are still suffering severe environmental and economic damage, there are encouraging signs. Some cities are learning how to better plan and more effectively manage the process of urban development, avoiding or alleviating environmental problems while realising the positive potentials of city growth and change. The SCP aims to support cities in finding—and managing—development paths that are more effectively fitted to their environmental opportunities and constraints.

There is a common approach that is shared by all SCP cities and that holds true across the full range of partner cities [9].

1. Central focus on development–environment interactions
2. Broad-based participation by public, private and community groups
3. Concern for inter-sectoral and inter-organisational aspects
4. Reliance on bottom-up and demand-led responses
5. Focus on process – problem-solving and getting things done
6. Emphasis on local capacity building

More recent initiatives in this field are promoted by various organisations. The Urban Low Emission Development Strategies (LEDS) project, funded by the European Commission and implemented by UN-Habitat and the International Council for Local Environmental Initiatives (ICLEI),* has the objective of enhancing the transition to low emission urban development in emerging economy countries. It offers selected local governments in Brazil, India, Indonesia and South Africa a comprehensive methodological framework (the GreenClimateCities methodology)† to integrate low-carbon strategies into all sectors of urban planning and development. Another initiative is the Cities and Climate Change Initiative (CCCI) that builds on UN-Habitat's long experience in sustainable urban development. The initiative helps counterparts to develop and implement pro-poor and innovative climate change policies and strategies. CCCI also is developing a suite

---

* ICLEI Local Governments for Sustainability is the world's leading association of more than 1000 metropolises, cities, urban regions and towns representing over 660 million people in 86 countries. ICLEI promotes local action for global sustainability and supports cities to become sustainable, resilient, resource efficient, biodiverse, low carbon; to build a smart infrastructure; and to develop an inclusive, green urban economy with the ultimate aim of achieving healthy and happy communities. Website: http://www.iclei.org
† ICLEI's GreenClimateCities programme offers a process to local governments integrating low emission alternatives into their planning processes and policies. It has a clear methodology, with guidance and/or tools provided for each step. For further information, refer to http://www.iclei.org/our-activities/our-agendas/low-carbon-city/gcc.html

of tools to support city leaders and practitioners in addressing the impact of climate change (adaptation) and to help to reduce greenhouse gas emissions (mitigation) [10].

2.3 Cities and Energy Consumption: The Microlevel

Within planning research, it is commonly assumed that the design and location of residential areas have important consequences for households' energy consumption for housing and transport. It is believed that physical planning and design make it possible to achieve a more sustainable consumption pattern. Mainly there are four distinct consumption categories: energy use for cooling/heating and operating the house; energy use for everyday travel; energy use for long leisure-time travel by plane; and energy use for long leisure-time travel by car.

2.3.1 Urban Pattern

In their study of the relationship between urban planning and energy consumption, Holden and Norland pose the question: Does the change of urban forms tend to reduce the frequency and length of journeys and, hence, energy consumption? To this day, the disagreement persists and the critiques against planning have many different forms, including [11]:

1. Claims that engine technology, taxes on gasoline and driving, and road pricing are more effective measures for reducing energy consumption than urban planning [12,13].
2. The assertion that socioeconomic and attitudinal characteristics of people are far more important determinants of travel behaviour than urban form. Critics in this matter emphasise that the importance of form is highly overestimated in empirical studies [14].
3. Casting doubt on the assumption that proximity to everyday services and workplace will contribute to reduced travel in a highly mobile society [15,16].
4. That the relationship between non-work travel, especially long leisure-time travel, and urban form has been neglected [17].
5. The assertion that travel preferences rather than urban form influence travel behaviour: People live in city centres because they prefer to travel less, and not that they travel less because they live in city centres (the 'self-selection bias') [13].

Even though these aspects should not be taken lightly, there seems to be overwhelming support in the literature for the idea that planning does matter in determining the level of energy consumption in urban areas. This view is based on theory and empirical studies advocating that planning is an important instrument for promoting sustainable development.

2.3.1.1 Compact versus Dispersed Development

When it comes to land-use characteristics that influence energy use for everyday transport, Næss concludes that the following characteristics are favourable for reducing energy use per capita: high population density for the city as a whole; high density within each residential area; centralised settlement within cities and towns (i.e. higher density in the inner part than on the fringe); centralised workplace location; low parking capacity at workplaces; decentralised concentration at the regional level; and a high population for each city [11,18].

The main principle in the compact city theory is high-density development close to or within the city core with a mixture of housing, workplaces and shops. This implies densely and concentrated housing development, which favours semidetached and multifamily housing. Under this theory, development of residential housing areas on (or beyond) the urban fringe, and single-family housing in particular, are banned. Furthermore, central, high-density development supports a number of other attributes that are favourable to sustainable energy use: low energy use for housing and everyday travel, efficient remote heating/cooling systems, proximity to a variety of workplaces and public and private services, as well as a highly developed public transport system.

The supporters of the compact city theory [19–24] believe that the compact city has environmental and energy advantages, as well as social benefits. The list of advantages is remarkably long, including a better environment, affordable public transport, the potential for improving the social mix and a higher quality of life [25]. However, the main justification for the compact city is that it results in the least energy-intensive activity pattern, thereby helping us cope with the issues of global warming. The supporters of the dispersed city suggest that the green city—that is, a more open type of urban structure, where buildings, fields and other green areas form a mosaic-like pattern [11,18].

The list of arguments against the compact city theory is even longer than the list in support of it and includes: that it rejects suburban and semi-rural living, neglects rural communities, affords less green and open space, increases congestion and segregation, reduces environmental quality and lessens the power for making local decisions [11,25].

However, until fairly recently, an international consensus favouring the compact city as a sustainable development approach has dominated the debate [26]. Although there has always been considerable scepticism, the concept of the compact city has been so dominant that it seems inconceivable that anyone would oppose the current tide of opinion towards promoting greater sustainable development and the compact city in particular [27].

Consequently, it is not surprising that the 'move towards the compact city is in the mainstream throughout Europe' [28, p. 275].

The disagreements between the compact city and dispersed city discourses can be summarised to a large extent as a debate about two issues—which form affords the greater energy efficiency, and which aspects of sustainable development are more important?

The relationship between urban form and energy efficiency—especially energy use for travel—is at the core of the sustainable urban form debate. During recent decades, there has been a multitude of empirical studies supporting the relative energy efficiency of the two urban forms. Boarnet and Crane worked through this literature and came to a rather surprising conclusion: 'Very little is known regarding how the built environment influences travel' [13, p. 4]. Although these authors were referring to the United States, we find the same scepticism in Europe. Williams et al. conclude that 'A great deal still needs to be learnt about the complexity of different forms and their impacts' [29, p. 335]. This includes the relationship between urban compactness and travel patterns. A possible relationship between the built form and long leisure-time travel by car and plane is a part of this new knowledge that has to be learned [11].

The possible impacts of urban forms are not limited to travel behaviour. The built form also influences social conditions, economic issues, environmental quality and ecology within the city [29]. All these aspects are also important parts of the sustainable development concept and therefore can be used as criteria for a discussion about sustainable urban form. It should come as no surprise that a study that has minimising energy consumption as an overall goal could easily reach different conclusions from those of a study that aims at using urban form to 'reduce the number of people exposed to fine particles' or to 'promote social equity'. In the end, it will be necessary to balance these impacts because sustainable urban form is ultimately about values [30].

The dispute between the two camps has led to the development of a number of middle positions, which try to combine the best aspects of the compact and the dispersed city discourses, while at the same time trying to avoid the disadvantages of each. Among such alternative middle positions are the urban village [31,32], 'New Urbanism', the sustainable urban matrix [33], transit-oriented development [13], smart growth [34] and decentralised concentration [35–37], and sustainable urbanism [38]. These alternatives all try to combine the energy efficiency gained from a compact urban form with the broader quality-of-life aspects gained from the dispersed city. Still, whether a specific urban form will be more energy efficient is an empirical question [11].

2.3.1.2 Density

Much of the concern with density in planning and other related fields has been over high urban density and its assumed negative effect on the quality of life of urban residents. The city has historically been perceived to be a place of overcrowding, noise, dirt, crime, poverty, disease and so forth [39–41]. The high density existing in cities during the early period of the Industrial Revolution was seen as one of the major culprits of poverty and disease. As a result, planning controls (in Canada and Great Britain, for example) usually specified maximum densities. The planning reaction was a strong movement towards lower density housing outside of the city. In the United States and Canada, this took the form of a move to the suburbs, but in Great Britain and Sweden, it resulted in garden cities [41,42]. Radberg describes the garden city movement as representing decentralised urban growth [39]. The assumption was that these relatively low-density residential areas would not suffer from the ills found in high-density cities and would offer a higher quality of life to residents [43].

More recently, there have been many second thoughts on, and strong criticisms of, these trends. Environmentalists express concern about the environmental implications of low density [44], and urbanists are concerned about the decline of the city [19,40] or of the community [45,46]. Questions about low densities also have been posed by those who are concerned about the efficient use of land and public services [40]; by feminists and researchers who argue that low-density suburbs are hostile to women's lives—especially employed women with children and single parents [47] and by sociologists who criticise the social homogeneity and the social segregation in these low-density areas [46,48]. There are some, of course, who mention all of these problems [43,49].

In 1994, a detailed set of principles were set out in *Sustainable Development: The UK Strategy* (Department of the Environment 1994a), which was subject to further revision in 1999 (*UK Government's Strategy for Sustainable Development 1999*). In this strategy, the land-use planning system was targeted for specific treatment and the foundations laid for more recent policy statements on car usage and urban layouts [50]:

> (24.20) Urban growth should be encouraged in the most sustainable settlement form. The density of towns is important. More compact urban development uses less land …
>
> The scope for reducing travel, especially by car, is dependent on the size, density of development, and range of services on offer …
>
> (24.26) Town and city centres must incorporate the best principles of urban design …

Indeed, the commission recommended that planning guidance should increasingly reflect the growing sustainable agenda and should become much more integrated with other public policy areas, notably economic policy [49].

Hitchcock [51] and Orchard [52] direct attention to the fact that, on the whole, the discussion about increasing density and reducing urban land consumption concentrates almost totally on residential densities. It neglects all of the other land uses that make up a city, even though these land uses represent a significant proportion of a city's total land area. If these non-residential land uses are not taken into account, the reduction in land consumption achieved by increasing residential density will not be as great as initially conceived because services and amenities will have to be augmented to accommodate the increased population [43].

There are a number of advantages from increasing densities, which can be summarised as follows [43]:

1. It can help protect agricultural land from urbanisation.

2. It results in less depletion of the natural resources needed for construction purposes [53].

3. Built forms that facilitate higher net densities may result in significant reductions in energy demands [15,54]. Energy use within buildings can be reduced by passive solar architecture, superior insulation and energy-saving technology [54] or by built forms with low-surface areas and combined heat/cooling and power systems [55]. Owens [15] notes that very different densities (ranging from 37 to 250 dwelling units per hectare) are attainable using combined heat and power systems, depending on discount rates and fuel prices.

4. Decreased pollution from vehicle exhausts can be achieved as a result of a decline in the use of cars, the mixing of land uses, the provision of efficient and accessible public transportation, and walking [15,54]. High densities have been found to be associated with lower gasoline consumption per capita [35,56]; however, this is a controversial issue [28,52,57].

5. Decreased emission of pollutants may result from energy-saving land-use plans and from energy-efficient buildings [53].

6. High density may result in a decrease in the total number of car trips [53]. Nasar found lower automobile dependency scores in high-versus low-density neighbourhoods [58]. These differences were greater for older people, women and households with no children. A decrease in the number of kilometres per trip may also result [54,59–61].

7. High density has been found to be related to a higher proportion of travel on public transit, to greater public transit service provision per person and to transit use by a higher proportion of workers [20,35]. Increased public transit use, in turn, may reduce pollution emissions (an environmental advantage).

8. High density enhances the opportunity to use public transportation because high density brings the development of public transportation systems to the thresholds of profitability and efficiency. The report prepared by Berridge Lewinberg Greenberg, Ltd. adopts several benchmarks for the relationship between residential density and transit use. It suggests that 17–75 dwelling units per net hectare are necessary to sustain significant transit use, and 150 dwelling units result in a modal split of different transportation types in which more than 50% are public transit [62].

9. As a result of an increase in transit use, traffic congestion in residential, work and commercial centres may decrease [62].

10. Public transit can be more energy efficient. Handy highlights that it is the set of choices correlated with density—not density itself—that shapes travel behaviour [63]. In this context, Bannister discusses the interaction between socioeconomic circumstances and people's propensity to travel with different frequencies, trip lengths and transportation modes [59]. Moreover, gender should be added to these intervening variables [64]. Self debates the effect that a change in density would make. He claims, for example, that a 50% increase in the density of Canberra, Australia, would produce only a modest increase in public transit use [65].

11. It offers more opportunities to walk or ride a bicycle to work, service and entertainment facilities [59,60].

12. High densities may result in economies of scale that facilitate the use of better quality and more attractive building materials [51].

13. It enables the use of a building complex as an element of the urban composition. It also allows for a variety of densities and types of construction in a given region. Variation in density and construction, in turn, makes the environment more interesting [51].

14. High-density development in the proximity of public transportation lines can decrease the demand for land located further from these lines [66].

15. High-density development as infill in existing areas can revitalise those areas and can reduce the pressure to develop open spaces [61].

On the other hand, urban density is a major factor that determines the urban ventilation conditions, as well as the urban temperature. Under given circumstances, an urban area with a high density of buildings can experience poor ventilation and strong heat island effect. In warm-humid regions, these features would lead to a high level of thermal stress of the inhabitants and to increased use of energy in air-conditioned buildings. However, it is also possible that a high-density urban area, obtained by a mixture of high and low buildings, could have better ventilation conditions than an area with lower density but with buildings of the same height. Closely spaced or high-rise buildings are also affected by the use of natural lighting, natural ventilation and solar energy. If not properly planned, energy for electric lighting and mechanical cooling/ventilation may be increased and application of solar energy systems will be greatly limited [67].

2.3.2 Land-Use Distribution and Home–Work Trip

The distribution of uses over the city plan is the main driving or restraining force of transportation. It is those trips made to different facilities that shape our daily activities, whether going to work, or using educational, health or other public services, or just for leisure. Housing location influences the distances to different types of facilities, and the spatial location of most of these facilities suggests that average travel distances will be shortest for inner-city residents. However, there are claims that high accessibility to different services might create an increased demand for transport. Moreover, opting for a wider range of jobs, shops and leisure activities might establish the need for more everyday travel.

2.3.2.1 New Urbanism and Transit-Oriented Development

In urban design literature, the development of what is loosely referred to as 'New Urbanism' applies a raft of sustainable objectives to new urban layouts. The evolution of this movement may be traced to the development of urban villages (in the UK) and sustainable growth management projects, also known as New Urbanism (in the United States), that have been 'directed toward creating an alternative to the typical car-dominated suburban sprawl that predominates on the fringe of virtually all western cities and towns' [68, p. 207].

The main design concept in New Urbanism is the creation of a 'module' or 'ped-shed' (walkable urban design and sustainable place making). It is made up of a walkable neighbourhood with a 400-m radius to shops, services and transport nodes in which the fabric creates a series of interconnected pedestrian friendly streets. It does not necessarily ban the private car; however, it serves to 'maximize interaction while minimizing the travel needed to do it' [68, p. 209]. The logic is that there will be a dramatic reduction in car parking provision. It decreases from the predominant post-war patterns of two or three spaces per dwelling to one space or less. Consequently, a link is established between reduced car parking standards and the design of mixed uses, small street blocks and interconnected streets [69]. At a more fundamental level, conventional Western post-war car parking layouts are challenged by the need to raise residential densities to make for greater land-use efficiencies [70] and to foster non-car-based trip generation where a provision of less than one space per dwelling is a desirable objective [71]. Morris and Kaufman acknowledged that this focus on New Urbanism will make a significant contribution to achieving more sustainable cities, yet they voiced concern that 'While the intentions and potential to re-shape cities and towns towards less car dependence is a strong thrust of many practitioners of new urbanism, the evidence of major gains on the ground is limited' [68, p. 208].

The two approaches, New Urbanism and transit-oriented development, do not target increasing densities—any increase in density that is achieved is basically a by-product of a minimal nature. The emphasis of the New Urbanism movement is on small towns. New urbanists envision towns or neighbourhoods that are compact, mixed use and pedestrian friendly [42]. The emphasis of transit-oriented development, whose principal proponent is Calthorpe [49,72], is to plan balanced, mixed-use areas with a simple cluster of housing, retail space and offices within a one-quarter mile walking radius of a light rail system. The motivation for transit-oriented development is to improve the ills brought about by dependence on the automobile and the mismatch that exists between old suburban patterns and the post-industrial culture. The goal is to preserve open space and reduce automobile traffic without necessarily increasing density. Calthorpe [73] defines average net residential densities of urban transit-oriented developments as 44 dwelling units per hectare, with densities of 62–123 units per hectare for up to three-story apartment buildings [43,72].

2.3.2.2 Long-Distance Leisure Time Travel: Compensatory Travel?

An important question that arises from looking at the wider issue of energy use and greenhouse gas emissions is whether, for certain income levels, reduced local everyday travel will be compensated for by increased long-distance leisure travel at other times. Is it the case that—for certain income levels—the sum of 'environmental vices' is constant and that households managing on a small everyday amount of transport create even heavier environmental strain through, for instance, weekend trips to a cottage or long-distance holiday trips by plane? In the professional debate, some [73] have claimed that people living in high-density, inner-city areas will, to a larger extent than their counterparts living in low-density areas, travel out of town on weekends—for instance, to a cottage—in order to compensate for the lack of access to a private garden. In addition to this 'hypothesis of compensation', others, including the Swedish mobility researcher Vilhelmson [74], have launched a 'hypothesis of opportunity', which asserts that the time and money people save due to shorter distance daily travel will probably be used for long-distance leisure-time travel [11].

A study conducted in Norway suggested that the total energy use decreases as density reaches a certain point, although the data indicate that the total energy use increases at higher density levels. This pattern is similar to a pattern in the relationship between energy use and city size found by a number of empirical studies of cities in Norway, Sweden and England [18]. According to these studies, up to a certain point, energy use per capita decreases as density increases, but thereafter energy use starts to increase. Thus, the advantages of 'megacities' or 'extreme density areas' seem to be outweighed by the advantages offered by more modest forms of urban compactness [11].

2.3.3 Road Network and Transportation Network

Transportation is the leading consumer of energy and fuel in the city. The spread of roads among extended urban areas has helped people easily commute within these vast areas, thus making distances irrelevant and promoting more and more dispersion.

2.3.3.1 Road Network

The road network connects the various parts of the city and connects the city with its surrounding context. Thus, it contributes to the efficiency of the city, the flow of people and goods, and consequently to the economic cycle.

However, the emphasis on road network design has created not so lively neighbourhoods. This was expressed by The Prince's Foundation when examining 'Sustainable Urban Extensions' in the UK, in which the problem is summarised thus:

> House builders place a high priority on complying with rules and guid-
> ance on highway engineering. They are anxious that their estates' street
> system should be adopted by the local authority with the minimum
> of negotiation and delay. Estates are consequently designed around
> road layouts based on loops, dead-end spines and cul-de-sacs, whose
> principal aim is to handle road traffic as efficiently and safely as possible.
> But as well as discouraging travel on foot or by bicycle, these 'roads first–
> houses second' designs can damage the harmonious grouping of houses
> and visual quality.... [75, p. 1]

2.3.3.2 Transportation

Because a compact city strategy is recommended to be adopted, an emphasis on the development of rail transport of great accessibility, safety, sustainability and environmental friendliness is the main target. In a study conducted in 25 megacities, the following was found [8]:

1. Transportation is seen as the single biggest infrastructure challenge by a large margin and is a key factor in city competitiveness.
2. With air pollution and congestion emerging as the two top environmental challenges, stakeholders predict a strong emphasis on mass transit solutions.
3. Cities are more likely to focus on incremental improvements to existing infrastructure, rather than on new systems.
4. Demand management is rarely mentioned as a major strategy for addressing the cities' transport problems.

2.3.3.3 Parking

Parking policy is commonly viewed as a complementary measure to reduce car use when combined with other initiatives. Sustainability seeks to establish less reliance than previously existed on private car usage—for example, by promoting compact urban development in areas well served by good public transport. Urban design policy promotes a departure from the 'roads first, houses later' philosophy (as dictated by many highway standards) to give precedence to the relationship among buildings rather than strict adherence to predetermined road design in new residential environments. A new design approach to car parking has emerged where there is a shift from the previously adopted orthodoxy of minimum standards to maximum ceilings (i.e. no more than one space per dwelling). Such a trend towards reduction of parking standards (and thus provision) is at variance with the projected growth in car ownership worldwide [50].

Research studies clearly demonstrate that a trade-off exists between relaxing current car parking standards and raising residential density [70,76]. Urban design commentators and practitioners increasingly lobby in favour of a 'car-free urbanism' in which the sustainable residential neighbourhood is based on radical rethinking of density and parking policy. The avoidance of inflexible standards will yield improved layouts, so that urban design can reclaim the city back from the car [77].

2.3.4 Buildings: Form, Height and Facade Treatment

Globally, buildings are responsible for approximately 40% of the total world annual energy consumption. Most of this energy is for the provision of lighting, heating, cooling and air conditioning [67].

One way of reducing building energy consumption is to design buildings that are more economical in their use of energy for heating, lighting, cooling, ventilation and hot water supply. Passive measures, particularly natural or hybrid ventilation rather than air conditioning, can dramatically reduce primary energy consumption. However, exploitation of renewable energy in buildings and agricultural greenhouses can also significantly contribute towards reducing dependency on fossil fuels. Therefore, promoting innovative renewable applications and reinforcing the renewable energy market will contribute to preservation of the ecosystem by reducing emissions at local and global levels. This will also contribute to the amelioration of environmental conditions by replacing conventional fuels with renewable energies that produce no air pollution or greenhouse gases. The provision of good indoor environmental quality while achieving energy and cost-efficient operation of the heating, ventilating and air-conditioning (HVAC) plants in buildings represents a multivariant problem. The comfort of building occupants is dependent on many environmental parameters including air speed, temperature, relative humidity and quality in addition to lighting and noise. The overall objective is to provide a high level of building performance (BP), which can be defined as indoor environmental quality (IEQ), energy efficiency (EE) and cost efficiency (CE).

IEQ is the perceived condition of comfort that building occupants experience due to the physical and psychological conditions to which they are exposed by their surroundings. The main physical parameters affecting IEQ are air speed, temperature, relative humidity and quality. EE is related to the provision of the desired environmental conditions while consuming the minimal quantity of energy. CE is the financial expenditure on energy relative to the level of environmental comfort and productivity that the building occupants attained. The overall CE can be increased by improving the IEQ and the EE of a building [67].

Urban planning has a considerable impact on the future EE of buildings, and planners lack useful tools to support their decisions. A study was made presenting a new method based on a genetic algorithm that is able to search for optimum urban forms in mid-latitude climates (35–50°). Here, more energy-efficient urban forms are defined as those that have high building absorptance in winter and low summer building absorptance. These forms can be designed by choosing among regular tridimensional building geometries with fixed floor space indices, which can be parameterised by adjusting the following variables: number of floors, building length ratio, grid azimuth and aspect ratio on both directions. The results obtained show that adequate urban planning, based on the consideration of the local radiation conditions as a function of latitude, may result in significantly better building thermal performance. In particular, it is concluded that the highest latitudes are more restrictive in terms of optimal solutions: pavilions (cross-sectional square blocks) are the best solutions for latitudes of 50° and terraces (blocks infinite in length) are preferred for 45°. For lower latitudes, all urban forms are possible. In terms of grid angle with the cardinal direction, it is concluded that the angle should stay between –15° and +15°, except for the latitude of 50° where it can ranges from –45° to +45°. For slab and terrace urban forms, the spacing between blocks in the north–south direction should be maximised, quantified by a building-height-to-street-width (aspect) ratio that decreases with the increase of latitude, ranging from 0.6 for a latitude of 35°, to 0.4 for a latitude of 45°. For pavilions, the north–south aspect ratio is independent of latitude and should stay close to 0.7. The pavilion is the urban form that allows for a larger number of floors [78].

Arguably, the most successful designs were in fact the simplest. Paying attention to orientation, plan and form can have far greater impact on energy performance than opting for elaborate solutions. However, a design strategy can fail when those responsible for specifying materials, for example, do not implement the passive solar strategy correctly. Similarly, cost-cutting exercises can seriously upset the effectiveness of a design strategy. Therefore, it is imperative that a designer fully informs key personnel, such as the quantity surveyor and client, about their design and be prepared to defend it. Therefore, the designer should have an adequate understanding of how the occupants or processes, such as ventilation, would function within the building. Thinking through such processes in isolation without reference to others can lead to conflicting strategies, which can have a detrimental impact upon performance. Likewise, if the design intent of the building is not communicated to its occupants, there is a risk that they will use it inappropriately, thus compromising its performance. Hence, the designer should communicate in simple terms the actions expected of the occupant to control the building [67].

2.3.5 Renewable Energy

Research into future alternatives has been and is still being conducted to solve today's complex problems such as the rising energy requirements of a rapidly and constantly growing world population and global environmental pollution. Therefore, options for a long-term and environmentally friendly energy supply have to be developed that lead to the use of renewable sources (water, sun, wind, biomass, geothermal, hydrogen) and fuel cells. Renewables could shield a nation from negative effects in the energy supply, pricing and related environmental concerns. For many years, hydrogen (for fuel cells) and the sun [for photovoltaics (PVs)] have been considered as likely and eventual substitutes for oil, gas, coal and uranium. They are the most abundant elements in the universe. The use of solar energy or PVs for everyday electricity needs has distinct advantages: avoiding consuming resources and degrading the environment through polluting emissions, oil spills and toxic by-products. A 1-kW PV system producing 150 kWh each month prevents 75 kg of fossil fuel from being mined. It avoids 150 kg of CO_2 from entering the atmosphere and keeps 473 L of water from being consumed. Electricity from fuel cells can be used in the same way as grid power—to run appliances and light bulbs and even to power cars because each gallon of gasoline produced and used in an internal combustion engine releases roughly 12 kg of CO_2, a greenhouse gas (GHG) that contributes to global warming [67].

Sunlight is not only inexhaustible but also only energy source that is completely non-polluting. The World Summit on Sustainable Development held in Johannesburg in 2002 committed itself to 'encourage and promote the development of renewable energy sources to accelerate the shift towards sustainable consumption and production'. Accordingly, it aimed at breaking the link between resource use and productivity. This can be achieved by the following:

1. Trying to ensure economic growth does not cause environmental pollution
2. Improving resource efficiency
3. Examining the whole life cycle of a product
4. Enabling consumers to receive more information on products and services
5. Examining how taxes, voluntary agreements, subsidies and regulation and information campaigns can best stimulate innovation and investment to provide cleaner technology

Until 2002, renewable energy contributed as much as 20% of the global energy supply worldwide [79]. More than two-thirds of this came from biomass use, mostly in developing countries, some of it unsustainable. Yet, the

potential for energy from sustainable technologies is huge. On the technological side, renewables have an obvious role to play. In general, there is no
problem in terms of the technical potential of renewables to deliver energy.
Moreover, there are very good opportunities for Renewable Energy Targets
(RETs) to play an important role in reducing emissions of GHGs into the
atmosphere, certainly far more than have been exploited so far. However,
there are still some technical issues to address in order to cope with the inter-
mittency of some renewables, particularly wind and solar. Yet, the biggest
problem with relying on renewables to deliver the necessary cuts in GHG
emissions is more to do with politics and policy issues than with technical
ones [79]. For example, the single most important step governments could
take to promote and increase the use of renewables is to improve access for
renewables to the energy market. This access to the market needs to be under
favourable conditions and, possibly, under favourable economic rates as well.
One move that could help, or at least justify, better market access would be
to acknowledge that there are environmental costs associated with other
energy supply options and that these costs are not currently internalised
within the market price of electricity or fuels [67].

Renewables are generally weather dependent and as such their likely
output can be predicted but not controlled. The only control possible is
to reduce the output below that available from the resource at any given
time. Therefore, to safeguard system stability and security, renewables
must be used in conjunction with other, controllable, generation and with
large-scale energy storage. There is a substantial cost associated with this
provision.

The recent REN21* report (2014) states that renewables have entered the
mainstream as we begin the Decade of Sustainable Energy for All (SE4ALL),†
mobilising towards universal access to modern energy services, improved
rates of EE and expanded use of renewable energy sources by 2030. In 2012,
renewable energy provided an estimated 19% of global final energy consumption, and it continued to grow in 2013. Of this total share in 2012,
modern renewables accounted for approximately 10%, with the remainder
(estimated at just over 9%) coming from traditional biomass. Heat energy
from modern renewable sources accounted for an estimated 4.2% of total
final energy use; hydropower made up about 3.8%, and an estimated 2% was
provided by power from wind, solar, geothermal and biomass, as well as
by biofuels. The combined modern and traditional renewable energy share

---

* REN21 is the global renewable energy policy multi-stakeholder network that connects a wide
  range of key actors. REN21's goal is to facilitate knowledge exchange, policy development
  and joint action towards a rapid global transition to renewable energy.
† The UN Secretary-General's initiative Sustainable Energy for All mobilises global action to
  achieve universal access to modern energy services, double the global rate of EE and double
  the share of renewable energy in the global energy mix by 2030. As the newly launched
  Decade for Sustainable Energy for All (2014–2024) unfolds, REN21 will work closely with the
  SE4ALL Initiative towards achieving its three objectives.

remained about level with 2011, even as the share of modern renewables increased. This is because the rapid growth in modern renewable energy is tempered by a slow migration away from traditional biomass and a continued rise in total global energy demand [80].

It is useful to codify all aspects of sustainability, thus ensuring that all factors are taken into account for each and every development proposal. Therefore, with the intention of promoting debate, the following considerations are proposed [67]:

1. Long-term availability of the energy source or fuel
2. Price stability of energy source or fuel
3. Acceptability or otherwise of by-products of the generation process
4. Grid services, particularly controllability of real and reactive power output
5. Technological stability, likelihood of rapid technical obsolescence
6. Knowledge base of applying the technology
7. Life of the installation—a dam may last more than 100 years, but a gas turbine probably will not
8. Maintenance requirement of the plant

However, the improved energy performance of cities from these kinds of initiatives is usually being outweighed by the increases in the use of fossil fuels by private transports that have occurred in recent years. This is the case all over the developed world, and particularly in the United States and Australia, where low-density urban sprawl has made it very difficult to introduce energy-efficient public transport systems. In cities with low-density sprawl where most people rely on private cars, it will be particularly important to introduce new transport propulsion such as fuel cell technology to make private transport and public transport less polluting and more energy efficient [81].

Really significant breakthroughs in urban EE and introduction of sustainable energy systems in cities will emerge only as a result of major changes in national energy policy. We have seen some significant breakthroughs in some countries, but far more needs to be done to transform our cities from fossil fuel junkies to sustainable, future-proof systems [81].

Thus, in order to reach a sustainable environment, a combination of policies and actions is essential (as proposed in Figure 2.3). These policies include sustainable energy production and renewable energy production as considerations for city planning regarding EE. When these policies are combined with trends in users' consumption patterns, a sustainable environment can be reached.

It is important to note that the know-how exists to decrease urban energy use by 50% or more without significantly affecting living standards, while

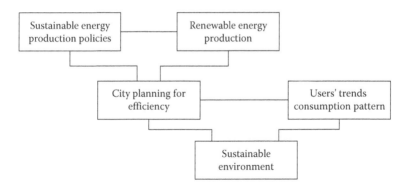

FIGURE 2.3
The relationship among policies for achieving a sustainable environment.

creating many new local jobs at the same time. There are some initiatives to improve the EE of cities—for example, The Cities for Climate Protection Program of the ICLEI. This is a performance-oriented campaign, started in 1993, offering a framework for local authorities to reduce global warming and waste gas emissions all over the world. This framework includes five performance milestones (which were implemented by 500 local governments participating in the campaign in 2004) [82]:

1. Conduct an energy and emissions inventory and forecast
2. Establish an emission target
3. Develop and obtain approval for the local action plan
4. Implement policies and measures
5. Monitor and verify results

2.4 City Consumption and City Impact

2.4.1 Ecological Footprint

In order to measure a city's impact versus its consumption, a more sophisticated analysis has been developed by Rees that can calculate a city's ecological footprint (EF) [83]. As previously mentioned in Chapter 1, it is based on an ecological understanding of how a city extracts food, water, energy and land from a bioregion (and beyond) and what ecosystem services it requires to absorb its wastes. The total resource use of a city is figured relative to its population, and the resulting calculation allows a per capita footprint of land to be compared to that of other cities [84].

The EF, as previously explained, translates consumption of various types into a common metric—the total area of productive land and water ecosystems required to produce the resources that the population consumes and to assimilate the wastes that the population produces, wherever on Earth that land and water may be located [85]. In calculating the footprint of nations or regions, the different bioproductivities of various land types are taken into account; this is achieved by incorporating equivalency factors, such that the calculated EF is expressed as standardised acres of world-average productivity. EFs quantify humans' overall impact on nature in relation to carrying capacity [86]. In 2000, the average global footprint was 6.25 acres per capita, but there were only 4.8 acres available per person based on the biologically productive area divided by the world population. Hence, we were in a deficit of 1.45 acres per person [86], depleting Earth's natural capital rather than living off nature's interest [87]. According to the Global Footprint Network for 2007, these numbers are 6.7 acres per person as EF per capita worldwide, while the biocapacity for Earth is only 4.4 acres per capita. Thus, the deficit has increased to 2.3 acres [88].

The concept of an EF is now firmly ensconced in the environmental literature and, despite its limitations [89–95], there is considerable support among researchers and environmentalists for the footprint as a clear, unambiguous indicator of human impact on nature that is easily applied [85,93,94,96].

One of the important linkages, that is not often drawn, is between EF, urban density and transport energy. Some commentators have criticised the use of per capita car use and per capita land use as confounding the statistics because population is in both denominators [97,98]. However, if the population factor is removed, then it is possible to look at whether land area (the direct footprint of a city) relates to transport [84].

One inherent weakness of using the EF is that it, like other inventory tools, is intended to measure impact. The EF is not designed look at cause and effect. However, where qualitative data provide insight into decision-making processes and choices, the EF becomes a useful tool for understanding the pathways to different outcomes. Also, the raw data assembled for its calculation could be used for specific questions of importance in planning practice.

In addition, the EF could be used by policy makers as part of the approval process for proposed developments. Rather than restricting development according to standard urban design codes, developments could be classified by a maximum EF. It would be up to developers and designers to plan communities that fall within the assigned EF. Rather than crippling innovation and creativity in urban design through legislation, a maximum assigned EF would foster new ideas and designs to tackle the sustainability challenge [87].

A study on the EF in EcoVillage at Ithaca, New York (United States), found that consumption, not built form, contributes most to the overall footprint;

therefore, the link between design and behaviour is of critical importance. The experiences at EcoVillage at Ithaca suggest that physical design may be a catalyst or facilitator of some changes in consumption, especially as they relate to utilities and possibly also to transportation, but no overall conclusion on the interaction between design and behaviour can be drawn from this study [87].

2.4.2 Sustainability Assessment

The EF helps when assessing development on the global scale, but on the local scale there is a need for a much more comprehensive tool. A key aspect to sustainability assessment is the assistance it provides to complex, controversial urban policy issues. One example is the density of cities and planned developments—a very controversial policy area in some urban settings. There is a strong global economic rationale for redeveloping car-dependent cities into focused centres and corridors to make better use of infrastructure at the scale required to provide such local services as public transport, shops and community services within walking distance. The lesser need for transport, the reduced urban sprawl and EF, the far greater opportunities for housing diversity, and other equity issues all provide additional justification at local and global levels. However, those local residents in the area where redevelopment is planned often perceive it as a threat to their local environment and social amenity. Sustainability assessment of such development can ensure there are real global economic, environmental and social benefits (often regional benefits but they may as well be global for many local people), but it can also ensure that developers include real local economic, social and environmental benefits. It can be used to ensure that there is a clear rationale for any development in terms of local environmental benefit (enhancing the local sense of place) and of local socioeconomic benefit (clear provision of better services). With these in place, the local and global issues can be seen to be resolved and a net benefit provided [84].

'Good' planning begins with an assessment of users' needs [99]. For example, transit stops are located in a way that is sensitive to demand. However, planning may also help to shape demand. Indeed, the very existence of planning reveals some general level of acceptance that land markets require guidance to ensure the provision of needs but in a sustainable manner. There are a number of arguments against sprawl; in some cases, suburban development has devoured many wetlands, with consequences for future water quality and supply [100,101], while in other parts of the world it has engulfed arable land. Auto-dependence and associated air pollution have severe implications for those with respiratory problems, and carbon dioxide emissions may contribute to climate change with unforeseeable consequences [87].

Cities will always be centres of consumerism. However, we can change the way they utilise resources. This can be done by conceptualising cities as sustainable ecotechnical systems, which requires converting their largely linear resource throughput into circular resource flows. EE, resource productivity, and urban and industrial ecology are key terms in this context [78].

2.5 Roles of Stakeholders in Planning for EE

2.5.1 Legislations and Laws Addressing Environmental Issues

In order to achieve more energy-efficient cities, where development is sustainable and environmentally responsive, laws and legislations should play a vital role. In 2000, the city of Barcelona introduced its mandatory 'solar ordinance'. All new housing, offices, restaurants, and public buildings have to install solar hot water systems if they use substantial amounts of hot water. Old buildings also have to be fitted with solar hot water systems when they are refurbished. Around the Mediterranean, use of solar hot water systems has become commonplace. In Japan, about 10% of all dwellings have their own solar hot water systems [81].

In German cities, solar PV panels are becoming commonplace, despite the country's relatively cloudy skies. This is primarily due to the German government's 'feed-in' legislation, which has fixed subsidies and favourable tariffs for owners of PV roofs. They used to be paid about 50 cents/kWh for selling their electricity back to the electricity grid, which is about four times the price paid to conventional electricity generators. The policy has led to a massive growth in demand for solar PV technology across the country. Similar policies have been introduced in Austria, France and Spain [81].

2.5.2 Governance

Better governance is a vital step towards better cities. With so many areas crying out for investment in better infrastructure, it is not surprising that funding emerges as a big issue for many stakeholders in a study survey done on megacities.* However, for those involved in city management, it is improvements to governance—rather than just money—that are the top

---

* A unique global research project undertaken by two independent research organisations, GlobeScan and MRC McLean Hazel, with the support of Siemens, the infrastructure provider. The goal of the project was to carry out research at the individual megacity level to gather objective data as well as perspectives from mayors, city administrators and other experts on local infrastructure challenges. The findings are based on an in-depth survey of over 500 megacity stakeholders, including elected officials, public- and private-sector employees, and influencers such as academics, NGOs and media. This survey was supplemented with extensive secondary research, to enable the team to shed light on the key challenges faced by global cities at various stages of development.

priority going forward. More than half of respondents with knowledge of urban management see improved planning as the priority for solving city problems, compared with only 12% that prioritise increased funding. In addition to more strategic planning, there is also a strong focus on managing infrastructure and services more efficiently. Both these goals will require cities to make the step from passive administration of existing services to a more active style of managing systems that focuses on improved efficiency and more measurable outcomes [8].

There is also a relationship between the scale of the environmental burdens and the appropriate roles of different levels of government. Some governance failures can be traced to a mismatch between the scale of the problem and the scale at which the response has been articulated. Local governance should not be expected to reduce carbon emissions voluntarily, although it can be a very appropriate level for driving local water and sanitation improvements. Global governance, on the other hand, is clearly needed to help develop institutional mechanisms to reduce contributions to global climate change, but it is inappropriate for developing institutional mechanisms for managing local water and sanitation systems. On the other hand, reducing local environmental burdens often requires support (or at least the absence of opposition) from global processes and institutions, while responses to global burdens often need to be rooted in local agency [102,103]. Moreover, cities and their needs are complex, and the traditional, departmentally organised approach to city governance needs to be rethought to enable more holistic solutions on the one hand and more responsiveness and accountability to citizens at a local level on the other [8].

The search for improved efficiency may require megacities to contract out the management of more services to the private sector. One of the more surprising findings in the survey is the fact that the main perceived advantage of private sector operation is improved efficiency (more than access to funding). Where cities do increase private sector involvement, they will need to create the right framework for success. There is a variety of models available, where ownership and operation of services can be shared. But when entering into partnerships with the private sector, the consequences must be well thought through, and success will require a 'context-sensitive' approach to privatisation, with overall control (and responsibility) resting with the public sector. If comprehensive governance models and efficient management structures are put in place, economic attractiveness, environmental protection, and quality of life for all citizens need not be contradictory goals [8].

Today, there is almost universal recognition in governments at all levels that it is essential to incorporate environmental considerations into urban planning and management. This provides significant benefits in every area of urban life, cutting across issues such as health, poverty, security, and economic development. Moreover, there is an essential call for better communication within the government and with other stakeholders involved in city planning and operation.

2.6 The Middle East Context

2.6.1 The Gulf Area

The Gulf area as an arid zone provides a challenge for architects and urban planners to build urban settlements that respond to the needs of inhabitants for climatic comfort and in the same time be sensitive to energy use and its consequences of climate change. This section reviews two approaches to tackle this issue and provide a climatic responsive built environment that is energy efficient.

2.6.1.1 A Return to Compact Cities

Over centuries, the climate in Arabia has become a major factor that shaped the daily life of local societies and, thus, the form of their cities. Old cities were characterised by their compactness, which stemmed from the need for protection from the harsh environment. Urban fabric has been dominated by the building masses, the limited number of enclosed public and outdoor spaces, and the inward-looking architecture. Besides its environmental utility, compactness also provided a physical support to the local community, reflecting its strong social structure and complex network of kinships. Nowadays, Gulf cities that are mostly shaped by the modern movement and American lifestyle are in complete negation with their past. An unprecedented sprawl effect is taking place all over the Gulf countries due to the heavy reliance on private transportation, high building technology, powerful air-conditioning systems and private housing [104].

A study by Ben-Hamouche on cities in Arabia recognised two historical shifts in the form of the city. The first one occurred during the industrialisation era from the old compact city to the modern dispersed city, and the second shift is expected to occur in the information age from the modern dispersed city back to the post-modern compact city through the combination of the concepts of sustainability and IT. He refers to the New Urbanism movement and its principles in his call for referral to compact cities as a remedy to the cancerous sprawl and suburbia [104].

Although this study claims that the information age will make the city more compact, due to the diminishing need for mechanical mobility, this increased accessibility might not lead to compactness. The sprawling may continue; only car usage might decrease but not necessarily increasing density.

2.6.1.2 Masdar City: Innovative Technologies [105]

As the geographical core of the Masdar sustainable energy initiative, Masdar City has been one of the elements to move forward the most quickly. The concept is simple but radical: zero-carbon and zero-waste. This involves a radical rethink of everything about the way that the city will function.

The 7 km² site selected is near the airport and about 17 km from the city of Abu Dhabi, and were it not in the desert, it would be classified as a 'greenfield' site. The fundamentals of the plan have been agreed, ground has been broken and phase one is underway. Initially, more than $300 million of procurement is in place, and an additional $1 billion was expected to be committed by the end of 2009. The city was due to be built in 7 years, at a total cost of $22 billion. The first $4 billion of this was coming from the Masdar Initiative, with the remaining $18 billion being raised through direct investments and other financial instruments. In 2013, the Abu Dhabi government has committed $15 billion to Masdar city. Moreover, more than $1 billion of equity has been invested across renewable energy projects with a total value of over $6.9 billion [106].

Sir Norman Foster, the British architect, is behind the design of the city, and detailed planning and preparation has been done by a range of international consultants and experts, including Pooran Desai from BioRegional, the UK consultancy WSP, Canada and United States-based CH2M Hill.

2.6.1.2.1 Building Design

Much of the design will adopt local, vernacular architectural principles, but this will also be mixed with a lot of cutting edge technology, some of it still in the experimental phase. The city will incorporate traditional medinas, souks and wind towers and will make use of open, public squares and narrow shaded walkways to connect homes, schools, restaurants and shops. The buildings themselves will then adopt a wide range of passive measures, and they should consume well under a quarter of the energy used by comparable buildings elsewhere in the region.

2.6.1.2.2 Transportation

There will be no cars in Masdar City—indeed, no internal combustion engines of any type. Instead, there will be a network of electric trams (a light rail transit or LRT system, which will also link to the planned Abu Dhabi LRT system) and smaller, 'personal rapid transit' vehicles, effectively an automatic, driverless system of electric taxis controlled by a central computer. These will be programmed so that, once occupied, the passenger has privacy and no other passenger can board along the route.

2.6.1.2.3 Renewable Energy

All the energy used in Masdar will be renewably generated, not only the electrical power but also that for heating, cooling and transport. The bulk of this is likely to come from one solar form or another. There will be power generation for a smart grid from solar thermal power and concentrating PV and also distributed PV throughout the city. The wind resource in Abu Dhabi is generally poor and will contribute little to the overall mix, but some geothermal and waste-to-energy, particularly from bio waste, are also likely to be significant contributors.

As well as providing a regional location, there are also numerous partnership opportunities for companies with technologies that may be used at Masdar. Among the energy technologies expected to be sourced are PV and solar thermal power generation (concentrating PV, parabolic trough and parabolic dish generation); advanced thermal waste treatment plants; geothermal systems that can be used for district cooling; and smart grid management systems. A range of other district cooling systems are also being considered, together with water desalination and grey-water treatment plants, and waste handling systems, including plasma and pyrolysis. More widely, procurement is also underway for IT systems, the transport infrastructure and facilities management and services.

The Arabic word 'Masdar' was chosen as the name of the project because one definition of the word is 'source'—in the sense of the root or spring from which things originate. For years, many good renewable energy projects have suffered through lack of access to sources of funding. The Masdar Initiative demonstrates that the combination of good projects and a plentiful source of funding can result in very rapid development of even the most ambitious plans. As such, it may also be a beacon for other places that are contemplating whether large-scale investment in renewables really can pay off.

2.6.2 Egypt

There is a growing awareness in Egypt about the change in climate since 1982 when that country established the Egyptian Environmental Affairs Agency (EEAA). Egypt was also one of the first Arab countries to sign the United Nations Framework Convention on Climate Change (UNFCCC) in 1992. Egypt has participated in and undertaken several actions that deal with climate change and environmental issues [107].

1. Ratification of the UNFCCC, the issuance of Law 4/1994 for the Protection of the Environment, and participation in various international workshops and conferences related to climate change to avoid having any international obligations on developing countries, including Egypt.

2. The Ministry of Electricity and Energy has established several projects in the field of new and renewable energy (wind, solar, hydro and bio) and has encouraged EE projects.

3. The Ministry of State for Environmental Affairs has established guidelines for the private sector to encourage investments in the field of clean energy projects, waste recycling and afforestation.

4. With the restructuring of the National Committee of Climate Change in 2007, as the coordinator on the national level related to climate change issues, by putting a visionary for needed policies and

strategies to deal with these issues, and by suggesting mechanisms required for implementation.

5. Maximising the benefit from Kyoto Protocol Mechanisms through implementing Clean Development Mechanism Projects.

An energy code for BP was established by the Housing and Building Research Center (HBRC) in 2006. It specifies the energy consumption of buildings according to their use and typology. This was an initiative to make buildings more energy efficient; however, this code has not been yet implemented.

2.6.2.1 Strategic Planning for Cities Programme

On the urban planning level, since 2008, there have been a lot of efforts made to upgrade more than 200 Egyptian cities. The programme, conducted by the GOPP, started strategic planning of cities under the auspices of the Ministry of Housing. A parallel programme is run and funded by the UN-Habitat concerning small cities (i.e. 25,000–50,000 inhabitants).

It is important to note that continuous urbanisation of rural areas in Egypt has created a unique case. One can find cities that are just villages in their structure, plan, network and physical and social infrastructure. These cities comprise most of the Egyptian urban context. This is primarily due to the way a city is defined by the government (according to population size). Typically, a city is defined as a settlement with more than 25,000 inhabitants. In other words, a village could become a city when its population exceeds this limit; however, it will still hold its rural characteristics, way of life, function and physical features.

In the context of the strategic planning of cities, three main sectors are studied: shelter and informal areas, infrastructure and local economic development. Three other sub-crosscutting sectors are investigated: local governance; environment; and poverty, women & vulnerability. Its main activities include preparing a city profile for the sectors investigated, a list of projects that are required by the city that represent its priorities, and a strategic plan with these projects situated in the appropriate locations that shows the road network, land uses and city limits. All processes are conducted with a participatory approach where all the city stakeholders are involved in the process of planning, prioritising and decision making. The final product is the strategic plan for the city.

The environment sector is mainly concerned with environmental hazards, pollution, noise and solid waste management and recycling. There is no mentioning of energy responsiveness or planning for maximising efficiency of energy use. However, these issues might be tackled depending on the environment consultant concept and the local context of the city under study.

Despite the programme's negligence related to energy responsive strategies, it provides a unique opportunity to really make our cities green and energy responsive. There are a number of ideas and actions that, if gathered and formulated into a strategy, could present a pioneer example within local contexts of the developing world.

In the progress of strategically planning the city of Ashmun in the governorate of Menoufia, Egypt, a number of ideas and requests were submitted by the local community that are energy responsive in essence (Figure 2.4) [108].

1. Preserving city boundaries with minimal increase just to accommodate future needed services and the preservation of agricultural land.
2. The need to build a ring road to increase transportation efficiency and decrease energy consumption and pollution.
3. Increasing the heights to double the street width (law specifies max building height = 1.5 street width) in structurally fit buildings to become four floors instead of only three floors. Thus dandifying existing urban areas instead of horizontal spread of the city.
4. Non-inclusion of sprawling houses in city limits in order to prevent or minimise future expansion on agricultural land.
5. Wise location of needed services, appropriate rates per capita of services and facilities, and their concentration in single location central to community.
6. Advocating mixed uses as commercial/residential uses.

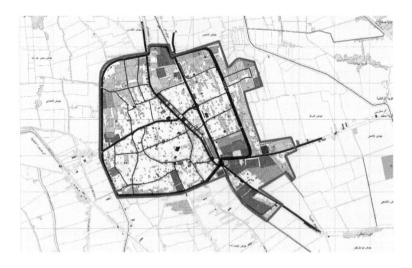

FIGURE 2.4
Keeping Ashmun city boundaries to the minimum and preserving surrounding agricultural land. (From Associated Consultants. (2008). *Strategic Planning for Ashmun City, Menoufia, Egypt*, Ministry of Housing, Utilities & Urban Communities and The General Organization of Physical Planning (GOPP), Cairo, Egypt.)

7. Locating workshops in a special area outside the residential mass, increasing efficiency of operation, management and transportation to other cities for further manufacturing.

8. The need to replace old deteriorated water supply asbestos pipes to prevent leakages and minimise health problems.

9. Better road network linkage with surrounding settlements for better and efficient transportation. Proposing a bridge to decrease travelling distance to neighbouring industrial zone (Sadat city). Thus providing jobs, preventing agricultural land loss and advocating more efficient industrial centres.

These ideas were required and enforced by the local community and the elected leaders, which shows awareness of the pressing issues of energy responsiveness and conventional resources depletion, despite the fact that a direct correlation to EE was not explicit. But this subtle concern can provide a solid base for more action to provide strategies for planning for EE in Egyptian cities.

2.6.2.2 Cairo

The megacity of Cairo is rated second worldwide for its pollution rate (Figure 2.5) [8]. There are approximately 18 million people living in the Greater Cairo region, which consists of five governorates. Expansion has caused many problems related to the environment, quality of life and infrastructure. Cairo is denser than many other cities because the law specifies an allowed maximum density of 150 person/feddan (357 persons/ha); it does not need to become any denser. However, there is a need to revise the current development strategies concerning the sprawling communities around the city. These gated communities are a replica of the American image of the perfect housing environment, where a villa exists on a private piece of land with a front lawn and a back garden. The building density is as low as 25% (downtown can reach 60%–70%) to accommodate for the extended open spaces. These communities were originally part of the green belt designated to surround Greater Cairo. However, there was a shift towards transforming it into dwelling areas, but with low densities, as an attempt to preserve the concept of the green belt.

Although there is a growing demand for these communities, they are far from environmentally friendly. The extension towards the new cities of Sheikh Zaied and 6th of October on one side of Cairo and the cities of Obour, Elshorouk and New Cairo on the other defies the original concept of establishing these as separate cities and transforms them into parts or districts of ever-growing Greater Cairo (see Figure 2.6).

This has really affected energy consumption trends, especially in increasing car dependency. The lack of an adequate transportation system that

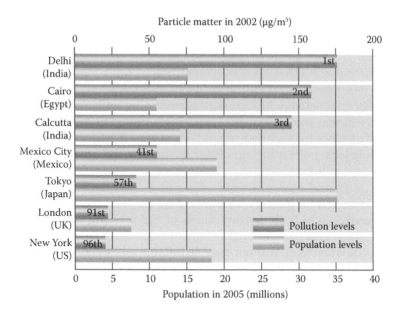

FIGURE 2.5
World's most polluted cities. (From The World Bank as cited in GlobeScan and MRC McLean
Hazel, *Megacity Challenges: A Stakeholder Perspective*, Siemens AG, Munich, 2007.)

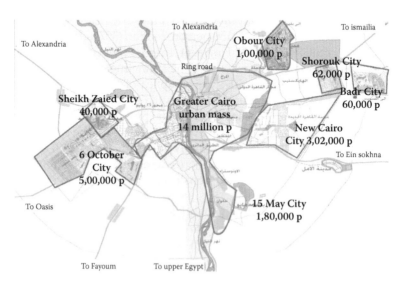

FIGURE 2.6
Greater Cairo with the surrounding new cities.

links all spread-out areas increases car dependency and fuel consumption. The home–work trip is becoming a daily nightmare resulting in congested traffic with a continuous peak hour.

Moreover, these gated communities and nearby new cities lack sub-centres that provide adequate services or businesses. Thus a trip to downtown Cairo is essential for obtaining services.

2.7 Conclusions

In November 2007, UN-Habitat held an Expert Group Meeting on 'Cities in Climate Change' in Nairobi, Kenya, bringing together participants from UN agencies, research institutions, local authorities and the private sector. The experts discussed the role of UN-Habitat regarding climate change and worked out basic elements for the agency's strategy on cities in climate change. The main outcomes of the Expert Group Meeting are that UN-Habitat has a clear role to play in dealing with climate change at the local level with a special focus on urban areas in developing countries. Furthermore, climate change should be regarded as a cross-cutting issue and integrated into UN-Habitat's existing initiatives and programmes [109].

The experts underlined the importance of immediate action—for example:

1. Launching of the Sustainable Urban Development Network (SUDNet) in 2008 for strengthening the performance of local governments to enhance climate change mitigation and adaptation measures in developing countries through existing and new partnerships
2. Promoting city-to-city cooperation
3. Conducting vulnerability assessments and risk mapping at the local level and providing guidelines for adaptive local planning
4. Collecting and sharing case studies on good practice
5. Developing mechanisms to assist cities in preventing land-use conflicts arising from relocation of human settlements
6. Assisting governments in translating National Adaptation Plans of Action to Local Adaptation Plans of Action together with adequate transfer of resources

This is in line with the experience gained from the Strategic Planning for Cities Program in Egypt because the actions required by the experts are what the programme in Egypt lacks on the broad level.

Much can be done in the developing countries. In the Egyptian context, there is a need for more awareness and action with regard to energy-efficient

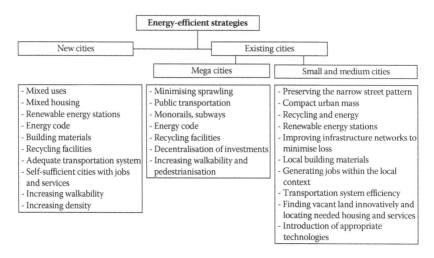

FIGURE 2.7
Energy efficiency strategies in cities.

strategies and to integrating them in urban planning. These strategies are summarised in Figure 2.7. On the personal level, it is recommended to try to use one's car less and separate his garbage as the former mayor of Curitiba, Brazil, advises [81].

References

1. Khalil, H. (2009). Energy efficiency strategies in urban planning of cities, *45th AIAA/ASME/SAE/ASEE Joint Propulsion Conference & Exhibit and 7th Annual International Energy Conversion Engineering Conference*, August 2–5, 2009, Denver, CO, paper No. AIAA 2009-4622.
2. World Energy Council. (2013). *World Energy Resources: 2013 Survey*, World Energy Council, London.
3. United Nations Environment Programme. *Activities: Urban—Energy for Cities*, Retrieved May 14, 2009, from http://www.unep.or.jp/Ietc/Activities/Urban/energy_city.asp
4. Chen, Q., Cheung, G., Hu, Y., Shen, Q., Tang, B., and Yeung, S. (2009). A system dynamics model for the sustainable land use planning and development, *Habitat International*, Vol. 33, pp. 15–25.
5. The United Nations Economic Commission for Africa (UNECA) and North Africa Office. (2003). The fight against desertification and drought in North Africa, *The Eighteenth Meeting of the Intergovernmental Committee of Experts*, United Nations Economic Commission for Africa, Tangiers, Morocco.
6. UN-Habitat. (2013). *State of the World's Cities 2012/2013: Prosperity of Cities*, Routledge, New York.

7. United Nations Human Settlements Programme. (2006). *The State of The World's Cities* Report 2006/2007: 30 Years of Shaping The Habitat Agenda, Earthscan, London, UK.

8. GlobeScan and MRC McLean Hazel. (2007). *Megacity Challenges: A Stakeholder Perspective*, Siemens AG, Munich.

9. The United Nations Centre for Human Settlements (UNCHS) and the United Nations Environment Programme (UNEP). (1999). *The SCP Source Book Series, Volume 5 Institutionalising the Environmental Planning and Management (EPM) Process*, UNCHS, Nairobi, Kenya.

10. UN-Habitat. (2012). *Evaluation Report 2/2012 Mid-Term Evaluation of the Cities and Climate Change Initiative*, UN-Habitat, Nairobi.

11. Holden, E. and Norland, I. T. (2005). Three challenges for the compact city as a sustainable urban form: Household consumption of energy and transport in eight residential areas in the greater Oslo Region, *Urban Studies*, Vol. 42, No. 12, pp. 2145–2166.

12. Gordon, P. and Richardson, H. W. (1989). Gasoline consumption and cities—A reply, *Journal of the American Planning Association*, Vol. 55, No. 3, pp. 342–345.

13. Boarnet, M. G. and Crane, R. (2001). *Travel by Design. The Influence of Urban Form on Travel*, Oxford University Press, New York.

14. Stead, D., Williams, J. and Titheridge, H. (2000). Land use, transport and people: Identifying the connections, In K. Williams, E. Burton, and M. Jenks, eds., *Achieving Sustainable Urban Form*, E & FN Spon, London, pp. 174–186.

15. Owens, S. (1992). Energy, environmental sustainability and land use planning, In M. J. Breheny, ed., *Sustainable Development and Urban Form*, Pion Ltd, London, pp. 79–105.

16. Simmonds, D. and Coombe, D. (2000). The transport implications of alternative urban forms, In K. Williams, E. Burton, and M. Jenks, eds., *Achieving Sustainable Urban Form*, E & FN Spon, London, pp. 115–123.

17. Titheridge, H., Hall, S., and Banister, D. (2000). Assessing the sustainability of urban development policies, In K. Williams, E. Burton, and M. Jenks, eds., *Achieving Sustainable Urban Form*, E & FN Spon, London, pp. 149–159.

18. NÆSS, P. (1997). *Fysisk Planlegging Og Energibruk* [Physical Planning and Energy use], Tano Aschehoug, Oslo.

19. Jacobs, J. (1961). *The Death and Life of Great American Cities: The Failure of Town Planning*, Random House, New York.

20. Newman, P. and Kenworthy, J. (1989). Gasoline consumption and cities: A comparison of US cities with a global survey, *Journal of the American Planning Association*, Vol. 55, No. 1, pp. 24–37.

21. Commission of the European Communities (CEC). (1990). *Green Paper on the Urban Environment*, European Commission, Brussels.

22. Elkin, T., Mclaren, D., and Hillman, M. (1991). *Reviving the City: Towards Sustainable Urban Development*, Friends of the Earth, London.

23. Sherlock, H. (1991). *Cities are Good for Us*, Paladin, London.

24. McLaren, D. (1992). Compact or dispersed? Dilution is no solution, *Built Environment*, Vol. 18, No. 4, pp. 268–284.

25. Frey, H. (1999). *Designing the City: Towards a More Sustainable Urban Form*, Spon Press, London.

26. Williams, K., Burton, E., and Jenks, M. (2000a). Achieving sustainable urban form: An introduction, In K. Williams, E. Burton, and M. Jenks, eds., *Achieving Sustainable Urban Form*, E & FN Spon, London, pp. 1–6.
27. Smyth, H. (1996). Running the gauntlet: A compact city within a doughnut of decay, In M. Jenks, E. Burton, and K. Williams, eds., *The Compact City. A Sustainable Urban Form?* E & FN Spon, London, pp. 101–113.
28. Jenks, M., Burton, E., and Williams, K., eds. (1996). *The Compact City: A Sustainable Urban Form?* E & FN Spon, London.
29. Williams, K., Burton, E. and Jenks, M. (2000b). Achieving sustainable urban form: Conclusions, In K. Williams, E. Burton, and M. Jenks, eds., *Achieving Sustainable Urban Form*, E & FN Spon, London, pp. 347–355.
30. Buxton, M. (2000). Energy, transport and urban form in Australia, In K. Williams, E. Burton, and M. Jenks, eds., *Achieving Sustainable Urban Form*, E & FN Spon, London, pp. 54–63.
31. Newman, P. and Kenworthy, J. R. (1999). *Sustainability and Cities: Overcoming Automobile Dependence*, Island Press, Washington, DC.
32. Thompson-Fawcett, M. (2000). The contribution of urban villages to sustainable development, In K. Williams, E. Burton, and M. Jenks, eds., *Achieving Sustainable Urban Form*, E & FN Spon, London, pp. 275–285.
33. Hasic, T. (2000). A sustainable urban matrix: Achieving sustainable urban form in residential buildings, In K. Williams, E. Burton, and M. Jenks, eds., *Achieving Sustainable Urban Form*, E & FN Spon, London, pp. 329–336.
34. Stoel, T. B., Jr. (1999). Reining in urban sprawl. *Environment*, Vol. 41, No. 4, pp. 6–11.
35. Breheny, M. (1996). Centrists, decentrists and compromisers: Views on the future of urban form, In K. Williams, E. Burton, and M. Jenks, eds., *The Compact City: A Sustainable Urban Form?* E & FN Spon, London, pp. 13–35.
36. Høyer, K. G. and Holden, E. (2003). Household consumption and ecological footprints in Norway: Does urban form matter? *Journal of Consumer Policy*, Vol. 26, pp. 327–349.
37. Holden, E. (2004). Ecological footprints and sustainable urban form, *Journal of Housing and the Built Environment*, Vol. 19, No. 1, pp. 91–109.
38. Farr, D. (2008). *Sustainable Urbanism: Urban Design with Nature*, John Wiley, Hoboken, NJ.
39. Radberg, J. (1998). Ebenezer Howard's dream: 100 years after. *Paper presented at the 44th International Federation of Housing and Planning World Congress*, September 13–17, Lisbon, Portugal.
40. Lehman and Associates. (1995). *Urban Density Study: General Report*, Office for the Greater Toronto Area, Toronto, Canada.
41. Gowling, D. and Penny, L. (1988). Urbanisation, planning and administration in the London region: Processes of a metropolitan culture. In H. van der Cammen, ed., *Four Metropolises in Western Europe*, Van Gorcum, Maastricht, The Netherlands, pp. 5–59.
42. Madanipour, A. (1996). *Design of Urban Space*, John Wiley, New York.
43. Churchman, A. (1999). Disentangling the concept of density, *Journal of Planning Literature*, Vol. 13, No. 4, pp. 389–411.
44. Van der Ryn, S. (1986). The suburban context. In S. Van der Ryn and P. Calthorpe, eds., *Sustainable Communities: A New Design Synthesis for Cities, Suburbs and Towns*, Sierra Club Books, San Francisco, CA.

45. Scully, V. (1994). The architecture of community, In P. Katz, ed., *The New Urbanism: Toward an Architecture of Community*, McGraw-Hill, New York, pp. 221–230.

46. Smyth, J. (1992). The economic power of sustainable development: Building the new American dream, In B. Walter, A. Lois, and R. Crenshaw, eds., *Sustainable Cities: Concepts and Strategies for Eco-City Development*, Eco-Home Media, Los Angeles, CA.

47. Churchman, A. (1993). A differentiated perspective on urban quality of life: Women, children and the elderly, In M. Bonnes, ed., *Perception and Evaluation of Environmental Quality*, UNESCO Programme on Man and Biosphere, Rome, pp. 165–178.

48. Shannon, G. and Cromley, E. (1985). Settlement and density patterns: Toward the 21st century, In J. Wohlwill and W. Van Vliet, eds., *Habitats for Children: The Impacts of Density*, Lawrence Erlbaum, Hillsdale, NJ, pp. 1–16.

49. Calthorpe, P. (1992). The pedestrian pocket: New strategies for suburban growth, In B. Walter, A. Lois, and R. Crenshaw, eds., *Sustainable Cities: Concepts and Strategies for Eco-City Development*, Eco-Home Media, Los Angeles, CA, pp. 27–35.

50. Stubbs, M. (2002). Car parking and residential development: Sustainability, design and planning policy, and public perceptions of parking provision, *Journal of Urban Design*, Vol. 7, No. 2, 2002, pp. 213–237.

51. Hitchcock, J. (1994). *A Primer on the Use of Density in Land Use Planning, Papers on Planning and Design no. 41*, Program in Planning, University of Toronto, Toronto, Canada.

52. Orchard, L. (1995). National urban policy in the 1990's, In Troy, P., ed., *Australian Cities: Issues, Strategies and Policies for Urban Australia in the 1990's*, University of Cambridge Press, Cambridge, pp. 65–86.

53. Breheny, M. (1992). The contradictions of compact city: A review, In M. Breheny, ed., *Sustainable Development and Urban Form*, Pion Ltd, London, pp. 138–159.

54. Stenhouse, D. (1992). Energy conservation benefits of high density mixed-use land development. In B. Walter et al., eds., *Sustainable Cities: Concepts and Strategies for Eco-City Development*, Eco-Home Media, Los Angeles, CA.

55. Rydin, Y. (1992). Environmental dimensions of residential development and the implications for local planning practice, *Journal of Environmental Planning and Management*, Vol. 35, No. 1, pp. 43–61.

56. Newman, P. and Kenworthy, J. (1989). *Cities and Automobile Dependence: An International Sourcebook*, Gower Technical Press, Brookfield, VT.

57. Gordon, P. and Richardson, H. (1997). Are compact cities a desirable planning goal? *Journal of the American Planning Association*, Vol. 63, No. 1, pp. 95–106.

58. Nasar, J. (1997). Neo-traditional development, auto-dependency and sense of community, In M. Amiel and J. Vischer, eds., *Space Design and Management for Place Making*, EDRA 28, EDRA, Edmond, OK.

59. Bannister, D. (1992). Energy use, transport and settlement patterns. In M. Breheny, ed., *Sustainable Development and Urban Form*, Pion Ltd, London.

60. Woodhull, J. (1992). How alternative forms of development can reduce traffic congestion. In B. Walter et al., eds., *Sustainable Cities: Concepts and Strategies for Eco-City Development*, Eco-Home Media, Los Angeles, CA.

61. Berridge Lewinberg Greenberg, Ltd. (1991). *Guidelines for the Reurbanisation of Metropolitan Toronto*, Municipality of Metropolitan Toronto Corporate Printing Services, Toronto, Canada.

62. Berridge Lewinberg Greenberg, Ltd. (1991). *Study of the Reurbanisation of Metropolitan Toronto*, Municipality of Metropolitan Toronto Corporate Printing Services, Toronto, Canada.
63. Handy, S. (1996). Understanding the link between urban form and nonwork travel behavior, *Journal of Planning Education and Research*, Vol. 15, No. 3, pp. 183–198.
64. Pickup, L. (1984). Women's gender-role and its influence on travel behavior, *Built Environment*, Vol. 10, No. 1, pp. 61–68.
65. Self, P. (1997). *Environmentalism and Cities. Newsletter Urban Research Program no. 32*, Research School of Social Sciences, Australian National University, Canberra.
66. Shireman, W. (1992). How to use the market to reduce sprawl, congestion and waste in our cities. In B. Walter et al., eds., *Sustainable Cities: Concepts and Strategies for Eco-City Development*, Eco-Home Media, Los Angeles, CA.
67. Omer, A. M. (2008). Energy, environment and sustainable development, *Renewable and Sustainable Energy Reviews*, Vol. 12, No. 9, pp. 2265–2300.
68. Morris, W. and Kaufman, J. (1998). The new urbanism: An introduction to the movement and its potential impact on travel demand with an outline of its application in Western Australia, *Urban Design International*, Vol. 3, No. 4, pp. 207–221.
69. Walsh, B. (1997). The right mix for urban living, *Urban Environment Today*, Vol. 14, February 20, pp. 8–9.
70. Llewelyn-Davies, South Bank University, Environment Trust Associates, and London Planning Advisory Committee. (1994). *The Quality of London's Residential Environment*, London Planning Advisory Committee, London.
71. McMullen, D. (2000). High densities—Key to meeting housing needs, *Urban Environment Today*, February 10, p. 9.
72. Calthorpe, P. (1993). *The Next American Metropolis*, Princeton Architectural Press, New York.
73. Kennedy, M. (1995). Ekologisk stadsplanering I Europa [Ecological urban planning in Europe], In *Den miljövänliga staden—en Utopi?* Rapport från en seminarserie, Miljöprosjekt Sankt Jörgen, Gøteborg.
74. Vilhelmson, B. (1990). *Va r dagliga rørlighet: om resandes utveckling, fördelning och gränser* [Our Daily Mobility: On the Development, Distribution and Limits of Travelling], TFB report 1990:16, The Swedish Transport Board, Stockholm.
75. The Prince's Foundation. (2000). *Sustainable Urban Extensions: Planned Through Design. A Collaborative Approach to Developing Sustainable Town Extensions through Enquiry by Design*, The Prince's Foundation, London, p. 1.
76. Oldfield King Planning Ltd. (1998). *Car Parking and Social Housing: National Planning and Housing Policy*, National Housing Federation, London.
77. Crilly, M. (1999). Novocastrian urbanism, *Urban Design Quarterly*, Vol. 72, p. 10.
78. Panão, M. O., Gonçalves, H. J., and Ferrão, P. M. (2008). Optimization of the urban building efficiency potential for mid-latitude climates using a genetic algorithm approach, *Renewable Energy*, Vol. 33, No. 5, pp. 887–896.
79. DEFRA. (2002). *Energy Resources. Sustainable Development and Environment*. Department of Environment, Food and Rural Affairs, DEFRA, Doncaster, UK.
80. REN21 Renewable Energy Policy Network for the 21st Century. (2014). *Renewables 2014, Global Status Report*, REN21 Secretariat, UNEP, Paris, France.

81. Girardet, H. (2004). *Urban Planning and Sustainable Energy: Theory and Practice*, Forum Barcelona, Retrieved May 12, 2009, from http://www.barcelona2004.org/www.barcelona2004.org/eng/banco_del_conocimiento/dialogos/fichac390.html

82. Gerardet, H. (2008). *Cities People Planet: Urban Development and Climate Change*, 2nd ed., John Wiley, West Sussex, England.

83. Rees, W. (1992). Ecological footprints and appropriated carrying capacity, *Environment & Urbanization*, Vol. 4, No. 2, pp. 121–130.

84. Newman, P. (2006). The environmental impact of cities, *Environment and Urbanization*, Vol. 18, No. 2, pp. 275–295.

85. Rees, W. (2000). Eco-footprint analysis: Merits and brickbats, *Ecological Economics*, Vol. 32, No. 3, pp. 371–374.

86. Chambers, N., Simmons, C., and Wackernagel, M. (2000). *Sharing Nature's Interest: Ecological Footprints as an Indicator of Sustainability*, Earthscan, London.

87. Andrey, J., Johnson, L. C., Moos, M., and Whitfield, J. (2006). Does design matter? The ecological footprint as a planning tool at the local level, *Journal of Urban Design*, Vol. 11, No. 2, pp. 195–224.

88. Ewing, B., Moore, D., Goldfinger, S., Oursler, A., Reed, A. and Wackernagel, M. (2010). *The Ecological Footprint Atlas 2010*, Global Footprint Network, Oakland, CA.

89. Gordon, P. and Richardson, H. (1998). Farmland preservation and ecological footprints: A critique, *Planning and Markets*, Vol. 1, No. 1, pp. 1–7.

90. Holmberg, J., Lundqvist, U., Robert, K. H., and Wackernagel, M. (1999). The ecological footprint from a systems perspective of sustainability, *International Journal of Sustainable Development and World Ecology*, Vol. 6, No. 1, pp. 17–33.

91. Van den Bergh, J. and Verbruggen, H. (1999). Spatial sustainability, trade and indicators: An evaluation of the ecological footprint, *Ecological Economics*, Vol. 29, pp. 61–72.

92. Deutsch, L., Jansson, A., Troell, M., Ronnback, P., Folke, C., and Kautsky, N. (2000). The ecological footprint: Communicating human dependence on nature's work, *Ecological Economics*, Vol. 32, No. 3, pp. 351–356.

93. Herendeen, R. (2000). Ecological footprint is a vivid indicator of indirect effects, *Ecological Economics*, Vol. 32, No. 3, pp. 357–358.

94. Moffatt, I. (2000). Ecological footprints and sustainable development, *Ecological Economics*, Vol. 32, No. 3, pp. 359–362.

95. Rapport, D. J. (2000). Ecological footprints and ecosystem health: Complementary approaches to a sustainable future, *Ecological Economics*, Vol. 32, No. 3, pp. 367–370.

96. Templet, P. (2000). Externalities, subsidies, and the ecological footprint: An empirical analysis, *Ecological Economics*, Vol. 32, No. 3, pp. 381–384.

97. Brindle, R. E. (1994). Lies, damned lies and 'automobile dependence'—Some hyperbolic reflections, *Australian Transport Research Forum*, Vol. 94, pp. 117–131.

98. Evill, B. (1995). Population, urban density, and fuel use: Eliminating spurious correlation, *Urban Policy and Research*, Vol. 13, No. 1, pp. 29–36.

99. Leung, H. L. (2003). *Land Use Planning Made Plain*, 2nd ed., University of Toronto Press, Toronto.

100. Draper, D. (1998). *Our Environment, A Canadian Perspective*, Thomson Nelson, Toronto.

101. Pollard, T. (2001). Greening the American dream? *Planning*, Vol. 67, No. 10, pp. 10–16.
102. Wilbanks, T. J. and Kates, R. W. (1999). Global change in local places: How scale matters. *Climatic Change*, Vol. 43, No. 3, pp. 601–628.
103. McGranahan, G. (2005). An overview of urban environmental burdens at three scales: Intra-urban, urban-regional, and global, special feature on the environmentally sustainable city, *International Review for Environmental Strategies*, Vol. 5, No. 2, pp. 335–356.
104. Ben-Hamouche, M. (2008). Climate, cities and sustainability in the Arabian region: Compactness as a new paradigm in urban design and planning, *Archnet-IJAR, International Journal of Architectural Research*, Vol. 2, No. 2, pp. 196–208.
105. Milford, E. (2009). Masdar city: A source of inspiration, *Renewable Energy World Magazine*, Vol. 12, No. 2, Retrieved from http://www.renewableenergyworld. com/rea/news/article/2009/04/masdar-city-a-source-of-inspiration, March 15, 2009.
106. Masdar's Fact sheet. (2014). Retrieved November 7, 2014, from Masdar: http:// www.masdar.ae/en/media/detail/masdar-fact-sheet-ver-1-jan-2013
107. Ministry of State for Environmental Affairs and Egyptian Environmental Affairs Agency. (2009). *Climate Change 2001–2009*, Retrieved May 3, 2009, from http:// www.eeaa.gov.eg
108. Associated Consultants. (2008). *Strategic Planning for Ashmun City, Menoufia, Egypt*, Ministry of Housing, Utilities and Urban Communities and General Organization of Physical Planning (GOPP), Cairo, Egypt.
109. Expert group meeting on cities in climate change, *Urban Environment Newsletter*, February 2008, Urban Environment Section, UN-Habitat, Nairobi, Kenya, in collaboration with UNEP, DTIE, Urban Environment Unit, Retrieved from http:// www.unhabitat.org/scp, February 15, 2009.

3

Energy-Efficient Urban Areas: Theories and Green Rating Systems*

3.1 Introduction

Sustainability is the governing paradigm of the 21st century. It has been the core of development debates for decades. However, expanding cities, especially in developing countries, are still lacking good guidance for coping with this paradigm. In these countries, the brown agenda dominates urban development; the green agenda is seen as a luxury and something to be considered in future generations. However, the current debate of quality of life (QOL) versus standard of living provides a strong base to support more environmentally responsive urban development. There is a growing awareness that providing a good QOL is a right of even the very vulnerable. This shift in paradigm advocates a more holistic approach to urban development. But question remains: What kind of urban planning theory or trend should be advocated to ensure a more sustainable development?

Currently, sustainable development is measured using a number of indicators and scales, of which ecological footprint and human development index are of vital importance. Moreover, there is a growing body of indices to rate urban agglomerations, whether on the macroscale of cities, such as the Green City Index, Low Carbon City Index and Comprehensive Assessment System for Building Environmental Efficiency (CASBEE) for cities, or the microscale of neighbourhoods and urban areas, such as Leadership in Energy and Environmental Design (LEED) for neighbourhoods. These indices tend to measure performance in a number of fields—namely energy and CO_2, land use, transport, waste, water, sanitation, air quality and environmental governance.

Cities are planned, managed and then are rated. Most of the recommendations and conclusion ratings fall within the field of urban management and urban governance. If the city has good urban governance, it will most likely be well managed. However, what about urban planning? Which trend

---

* Most of this chapter is derived from Khalil [1].

of planning facilitates good urban management and provides a stable groundwork for good governance to stand on?

In contrast to the aforementioned efforts to rate the sustainability of urban agglomerations, there are a number of theories that claim to create or promote sustainable urbanism; these are new urbanism (NU), smart growth, transit-oriented development (TOD) and sustainable urbanism. However, little evidence exists to support the link between following a certain theory and achieving precise progress on the sustainability charts. Developing countries, especially in Africa, are urbanising at a very fast rate. What about the cities to be built? What are the concepts that should govern urban planning of these new settlements? What lessons could be learned from rating existing cities to plan more successfully sustainable cities?

This chapter attempts to study this link and to analyse the relationship between adopting sustainable urbanism theories and scoring high on the green indices with a focus on energy efficiency. The chapter will analyse the urban pattern of highly ranked cities and correlate it to sustainable urbanism theories. It includes a matrix of relations highlighting the most important aspects or concepts that should be adopted to improve the energy efficiency score of a city. It further applies this matrix to one of Egypt's new cities to provide a forward vision for improving QOL within the Egyptian context.

3.2 Quality of Life Is a Right

QOL has been the domain of development discourse for the past decade. It has been widely recognised that measuring progress in terms of gross domestic product (GDP) is not sufficient. According to the ecological economist Costanza [2], although QOL has long been an explicit or implicit policy goal, adequate definition and measurement have been elusive. Diverse 'objective' and 'subjective' indicators across a range of disciplines and scales, and recent work on subjective well-being surveys and the psychology of happiness, have spurred renewed interest.

It is widely accepted now that seeking better QOL should be the ultimate goal of development plans and not just concentrating on economic progress. Currently, there is a shift away from depending on the GDP as a measure of material well-being. Consequently, there are a number of indices proposed and used by different organisations to score and rate cities and countries according to their QOL. These various indices were discussed in Chapter 1: Quality of Living by Mercer [3], Quality of Life Index by the Economist Intelligence Unit (EIU) [4], nations' ranking according to QOL [5], Your Better Life Index by the Organisation for Economic Cooperation and Development (OECD) [6], and City Prosperity Index by UN-Habitat.

All these indices attempt to measure how much QOL a person enjoys with a continuous emphasis that it is multifaceted. After years of relying on the human development index (HDI) as the main measure of progress, the shift of UN agencies, especially UN-Habitat, to adopting QOL as a measure of progress recognises that it should be a right and not a luxury.

3.3 Measuring Sustainable Development

It is essential to measure how cities are performing with respect to sustainable development. Thus, the need for combining indices or for an integrative tool is crucial to bring together environmental and human issues. According to Newman, this approach has a much more positive agenda because it encourages a much more pro-urban process and has the potential to drive an integrated policy agenda [7]. Sustainability assessment emerged mainly from the need to resolve the tension between ecologists, who saw most development as essentially negative, and those working for social justice (especially in low- and middle-income countries), who saw development as crucial to meeting human needs. Thus these assessments can help in shaping a development approach that would allow present and future generations to benefit economically, socially and environmentally.

According to Moos et al., there are four general approaches to assess the sustainability of different types of development [8,9]. The first focuses on the extent to which resources and ecological functions are protected from development, thus comparing how well two designs conserve the most ecologically valuable land. The theoretical foundation was led by environmental planning theorists and practitioners such as McHarg [10] and Hough [11], and many of these principles are also evident in what is commonly known as ecosystem planning [12,13]. However, this approach is limited in its ability to deal with the issue of sprawl and the implications of sprawl for transportation patterns and resource consumption more generally.

A second approach depends on ecosystem indicators, relying species diversity or forest health to measure how human activities impact the environment. This approach focuses on assessing changes over time, although resultant data also provide an opportunity to compare different developments or regions. Its shortcoming is the difficulty of translating the indicators into prescriptive information for planners and developers.

The third type of assessment measures the performance of a certain project against a set of established criteria. The U.S. Green Building Council's LEED, the British Building Research Establishment Environmental Assessment Method for buildings (BREEAM), and the Japanese CASBEE certification process have all created sustainability criteria for buildings and other larger scales.

The fourth type of assessment combines various environmental impacts into a common metric to facilitate comparisons over time and space. Both development attributes and consumptive patterns of residents or businesses occupying the space [14] are considered to provide some prescriptive information on how to improve the built form (e.g. ecological footprint). However, in order to be comprehensive, data assembly and the translation of various activities into a common metric become challenging [9].

Chapter 1 has highlighted the Human Development Initiative—where the ecological footprint and HDI have been combined. The initiative targeted how countries in the quest to achieve higher HDI consume (or not) their environmental resources and whether current development actions are sensitive to the country's ecological capacity or not.

3.3.1 Indices to Rate Urban Agglomerations

Globally, tools for evaluating cities are not as widely available as they are for buildings. There is a growing demand for tools to evaluate measures and activities on the scale of a city or a society as a whole. Currently, there are a number of indices to rate urban agglomerations varying in scale and thus the criteria or indicators they monitor or rate. Among these indices related to the urban scale are CASBEE for urban development and CASBEE for cities in Japan, LEED for neighbourhoods in the United States, and the Green City Index developed by the EIU and Siemens. Other indices related to assessing the sustainability of settlements that are tailored to local circumstances include megacity sustainability indicators in Brazil [15], the Sustainability Cities Index rating the biggest 20 cities in the United Kingdom [16] and the Freiburg Charter for Sustainable Urbanism [17].

3.3.1.1 *Comprehensive Assessment System for Building Environmental Efficiency (CASBEE)*

CASBEE is a tool developed in Japan under the leadership of the Ministry of Land, Infrastructure and Transport. It has several scales, such as for housing and offices, and in recent years, the ministry has been working on developing a low-carbon version of CASBEE for cities. Included in the process are the CASBEE tool for housing and other buildings as well as a tool for urban development and a tool for evaluating a city as a whole. The building and urban development tools were completed in 2007, and a third comprehensive tool was released in 2011 [18]. One fundamental principle guiding this effort is the need to reduce environmental loads and to create a low-carbon society while also improving the QOL. Clear, simple and comprehensive evaluation results should also be presented in a visual format, which is very important in obtaining consensus from the general public [19].

One example of the application of the CASBEE tool for urban development is Harumi Triton Square, a large-scale urban redevelopment area located about 1 km from Tokyo Station. It was evaluated as 'A with four stars' on a ratings scale or one to five stars [20].

Completed city environmental performance assessments would provide strong support for environmental measures within each municipality, and they would also function as a tool for comparing the environmental efficiencies of various measures with other cities. This would attract more attention from citizens and enhance their sense of belonging to their local communities. It is also expected to help motivate the community itself by introducing a competitive consciousness among cities [19].

3.3.1.2 LEED for Neighbourhood Development

In 2000, the U.S. Green Building Council (USGBC) launched its rating system, LEED, a set of standards for green buildings. The LEED standard combines prerequisites with optional credits that count towards an overall score. The levels of performance start at certified on the low end and platinum on the high end.

However, LEED has faced some criticism. Farr criticised its shortcomings because it is building-centred and places low value on a project's location and context, particularly concerning auto-dependency. In 2002, the USGBC Board of Directors inaugurated the LEED for Neighbourhood Development (LEED-ND) rating system in partnership with the Congress of New Urbanism (CNU) and the National Resource Defense Council. It increased the weighting given to land use and transportation concerns. In 2005, the USGBC board modified its mission to address both buildings and community [21]. The resulting rating system integrates the principles of smart growth, NU and green building into the first national standard for green neighbourhood development.

As stated by the LEED-ND guidebook, each of these three principles is an integral part of the LEED-ND rating system. Although each has a unique perspective, they reinforce each other by providing a set of criteria designed to help developers and designers envision sustainable communities in terms of where they are located, how they are designed and how they perform. It places emphasis on the site selection, design and construction elements that bring buildings and infrastructure together into a neighbourhood and relate it to its landscape as well as to its local and regional context [22].

Scale of eligible projects may range from whole neighbourhoods, portions of neighbourhoods, or multiple neighbourhoods, with no absolute minimum or maximum size limit. However, the recommended size range is at least two buildings up to 320 acres. Moreover, it is recommended that the majority (more than 50%) of planned project square footage consists of new construction or substantial renovation to ensure that the project has enough flexibility to meet a substantial number of requirements in the LEED-ND rating system.

Although projects may contain only a single use, typically a mix of uses will provide the most amenities to residents and workers and will enable people to drive less and safely walk or bike more. Small infill projects that are single use but that complement existing neighbouring uses, such as a new affordable housing infill development in a neighbourhood that is already well served by retail and commercial uses, are also good candidates for certification.

It is vital to note that LEED-ND is not intended to certify an entire town, county or city. Nor was it designed to rate existing neighbourhoods with no new development. Its main goal is to be applied during the design of new developments or during redevelopment and major retrofits. In addition, it can be used to analyse whether existing development regulations—such as zoning codes, development standards, landscape requirements, building codes, or comprehensive plans—are 'friendly' or are forming barriers to sustainable developments. LEED-ND can also be used as a tool to meet greenhouse gas reduction targets and as a 'best practice' manual for new development.

3.3.1.3 *Green City Index*

The Green City Index, as mentioned in Chapter 1, measures the current environmental performance of major cities on different continents as well as their commitment to reducing their future environmental impact by way of on-going initiatives and objectives. The EIU, in cooperation with Siemens, developed the methodology of the index.

The index scores cities across eight categories: CO_2 emissions, energy, buildings, transport, water, waste and land use, air quality, and environmental governance; and it has 30 individual indicators. Sixteen of the index's 30 indicators are derived from quantitative data and aim to measure how a city is currently performing. The remaining 14 indicators are qualitative assessments of cities' aspirations or ambitions [23]. Currently, this index covers areas of Europe, Latin America, Asia, the United States and Canada, Germany and Africa [24].

Despite fuelling some debates regarding the choice of certain indicators and the neglect of others, this index offers a tool to enhance understanding and to highlight needed interventions for improving the environmental performance of a particular city.

3.4 Sustainable Urbanism Theories

In the quest for forming a theory for sustainable urban planning, the debate between going compact or dispersed has dominated the philosophical and empirical research. The main principle in the compact city theory is high-density development close to or within the city core with a mixture of housing, workplaces and shops [25]. This debate, where the supporters of each growth pattern claim to attain sustainability in a more comprehensive way, was discussed in Chapter 2. The supporters of the compact city theory believe that the compact city has environmental and energy advantages as well as social benefits. On the other hand, as stated by Næss, the supporters of the dispersed city support the idea of the green city—that is, a more open type of urban structure, where buildings, fields and other green areas form a mosaic-like pattern [25,26]. However, dispersed development that embraces nature into the urban context has other impacts that are not at all sustainable. Sprawling land development has been consuming most of the American countryside at an alarming rate. Sprawl is defined as development that is dispersed, auto-dependent, single use, and where it is impossible to walk to your daily needs. There is a growing general awareness that low-density residential development threatens farmland and open space, raises public service costs, encourages people and wealth to leave central cities, creates serious traffic congestion, and degrades the environment and our QOL [27].

The disagreements between the compact city and dispersed city discourses can, to a large extent, be summarised as a debate about two issues: Which form affords the greater energy efficiency, and which aspects of sustainable development are more important? [28].

As discussed in Chapter 2, a number of in-between ideas have emerged that try to achieve energy efficiency and provide broader QOL. Examples of such theories are the urban village, NU, the sustainable urban matrix, TOD, smart growth, decentralised concentration and sustainable urbanism.

3.4.1 NU and Smart Growth

One of the strongest theories advocating for compactness and objecting to sprawling is that of NU. The CNU was founded by six architects: Peter Calthorpe, Andres Duany, Elizabeth Moule, Elizabeth Plater-Zyberk, Stephanos Polyzoides and Daniel Solomon in 1993 and were assisted in the coordination of their effort by Peter Katz, who became the first Executive Director of CNU. The main goal of the movement and subsequent theory and charters is to promote traditional urbanism as an antidote to conventional sprawl and to write a charter that would rebut Congrès Internationaux d'Architecture Moderne (CIAM) and its Athens Charter and would serve as a governing document for this reform movement.

Farr states that the greatest strength of the CNU has been its design excellence and rhetorical mastery in communicating the vocabulary of urbanism as it related to clients' projects. It has excelled in creating mixed-use neighbourhood developments and transit villages, featuring town centres, fine-grained walkable street grids, and a highly diverse ensemble of traditional buildings and architectural styles. He adds that the two new urbanist innovations, the urban-rural transect and the smart code, both developed by Andres Duany, have the potential to reshape the urban development process. The urban–rural transect applies the ecological framework of the natural transect to describe human settlements across a spectrum of intensity ranging from wilderness to dense urban centres. The smart code is a transect-based, form-based code that seeks to replace existing zoning codes with new codes that are simple and clear and that can be calibrated locally [21].

The principles of urbanism can be applied increasingly to projects at the full range of scales from a single building to an entire community. These principles are as follows [29,30]:

1. Walkability: Implying that (a) most things are within a 10-minute walk of home and work; (b) pedestrian-friendly street design (buildings close to street; porches, windows, and doors; tree-lined streets; on street parking; hidden parking lots; garages in rear lane; narrow, slow speed streets); and (c) pedestrian streets free of cars in special cases.

2. Connectivity: Having (a) an interconnected street grid network that disperses traffic and eases walking; (b) a hierarchy of narrow streets, boulevards, and alleys; and (c) a high-quality pedestrian network and public realm that makes walking pleasurable.

3. Mixed-use and diversity: Having (a) a mix of shops, offices, apartments and homes on site; mixed-use within neighbourhoods, within blocks and within buildings; and (b) diversity of people—of ages, income levels, cultures and races.

4. Mixed housing: With a range of types, sizes and prices in closer proximity.

5. Quality architecture and urban design that emphasise beauty, aesthetics and human comfort; create a sense of place; and advocate special placement of civic uses and sites within the community. Moreover, the human scale architecture and beautiful surroundings nourish the human spirit.

6. Traditional neighbourhood structure, where there is (a) a discernible centre and edge with public space at the centre; (b) importance of quality public realm; public open space designed as civic art; (c) a range of uses and densities within a 10-minute walk; and (d) transect planning—highest densities at town centre and progressively less dense towards the edge. This urban-to-rural transect hierarchy has appropriate building and street types for each area along the continuum.

7. Increased density where more buildings, residences, shops and services are closer together for ease of walking, enabling a more efficient use of services and resources, and creating a more convenient, enjoyable place to live.

8. Green transportation including (a) a network of high-quality trains connecting cities, towns and neighbourhoods and (b) a pedestrian-friendly design that encourages a greater use of bicycles, rollerblades, scooters and walking as daily transportation.

9. Sustainability, insuring (a) minimal environmental impact of development and its operations; (b) ecofriendly technologies, respect for ecology, and value of natural systems; (c) energy efficiency, less use of finite fuels; (d) more walking, less driving; and (e) more local production.

10. QOL that results from the previous aspects, where places are created that enrich, uplift and inspire the human spirit.

Smart growth initiatives identify the relationship between development patterns and QOL by implementing new policies and practices promoting better housing, transportation, economic development, and preservation of environmental quality. It traces its roots back to the 1970s; however, it was consolidated when Harriet Tregoning, then Director of Development, Community, and Environment at the U.S. Environment Protection Agency, developed the ten principles of smart growth in 1996. At that time, many environmentalists were simply antigrowth, viewing all development—without distinction—as hostile to the environment. These principles succeeded in uniting a decentralised grassroots movement of local and regional citizen activists and municipal leaders under the Smart Growth banner [21].

The ten principles of smart growth are: (1) mix land uses; (2) take advantage of compact building design; (3) create housing opportunities and choices for a range of household types, family sizes and incomes; (4) create walkable neighbourhoods; (5) foster distinctive, attractive communities with a strong sense of place; (6) preserve open space, farmland, natural beauty and critical environmental areas; (7) reinvest in and strengthen existing communities and achieve more balanced regional development; (8) provide a variety of transportation choices; (9) make development decisions predictable, fair and cost-effective; and (10) encourage citizen and stakeholder participation in development decisions [27].

3.4.2 Transit-Oriented Development

Also known as transit-oriented design, or TOD, this is the creation of compact, walkable communities centred around high-quality train systems, making it possible to enjoy a higher QOL without complete dependence on a car for mobility and survival. It is a major solution to the serious and growing problems of peak oil and global warming by creating dense, walkable communities connected to a train line that greatly reduces the need for driving and the burning of fossil fuels.

Components of TOD include (a) a walkable design with pedestrian as the highest priority; (b) a train station as the prominent feature of the town centre with a regional node containing a mixture of uses in close proximity including office, residential, retail and civic uses; (c) high-density, high-quality development within a 10-minute walk circle surrounding the train station; (d) collector support transit systems including trolleys, streetcars, light rail, buses and so forth; (e) easy use of bicycles, scooters and rollerblades as daily support transportation systems; and (f) reduced and managed parking inside a 10-minute walk circle around the town centre/train station [31].

3.4.3 Sustainable Urbanism

In a more comprehensive movement, Farr established the concept of 'sustainable urbanism'. He states its basic tenets as walkable and transit-served urbanism integrated with high-performance buildings and high-performance infrastructure. Sustainable urbanism emphasises that the personal appeal and societal benefits of neighbourhood living—meeting daily needs on foot—are greatest in neighbourhoods that integrate five attributes: definition (defined centre and edge), compactness (increasing sustainable effectiveness), completeness (daily and lifelong utility), connectedness (integrating transportation and land use), and biophilia (human access to nature) [21].

According to Khalil, guidelines for sustainable urban development can be formulated using a number of criteria: (a) Defined neighbourhood with a defined centre and edge comprising a traditional neighbourhood structure and advocating high-quality architecture and urban design; (b) compactness for an increased sustainable effectiveness through advocating walkability, connectivity, increased density and compact building design; (c) completeness with daily and lifelong utilities including diversity of mixed uses, mixed housing and fostering distinctive, attractive communities with a strong sense of place; (d) connectedness with integrating transportation and land use, green transportation and a variety of transportation choices; (e) enhancing the QOL by preserving open space, farmland, natural beauty and critical environmental areas and a more balanced regional development, promoting human access to nature; and (f) such sustainable urban development should encourage citizen and stakeholder participation in the development decisions [30].

3.5 Best Practices in Energy Efficiency and Sustainable Urbanism Principles

This section analyses the relationship between scoring high in energy efficiency on the Green City Index and adopting the relevant principles of sustainable urbanism. As shown in Table 1.1 in Chapter 1, indicators concerning energy efficiency have been extracted from the Green City Index. These indicators will be used to study the relevance to sustainable urbanism principles for the highest-ranking city. In the European Green Index, Copenhagen scores as Europe's greenest city. Successive governments at the national and municipal levels have strongly supported the promotion of sustainable development and have been taking environmental issues and sustainable energy seriously since the oil shock of the 1970s [23]. In Latin America, Curitiba scores highest as Latin America's greenest metropolis, continuing to live up to its reputation for sustainable urban planning [32]. It has been pursuing a long-term strategy since the 1960s to control urban sprawl and to plan and manage its transportation systems. In the African index, Johannesburg was chosen because it ranks above average overall, along with five other cities. It is particularly strong in energy and CO_2, land use, transport, air quality and environmental governance, ranking above average in each category [33].

The previous analysis of best practices in Europe, Latin America, and Africa (as specified in Table 3.1 [23,32,33]) has shown the relationship between scoring high on energy efficiency indicators and applying or fulfilling principles of sustainable urbanism. For decades, Copenhagen and Curitiba have been famous for their adoption of sustainable urbanism principles. Therefore, it is clear that urban planning is the major tool to achieve eco-efficiency and not limit actions to urban management and proposing policies. Johannesburg, as with most African cities, is still expanding and thus needs to develop a comprehensive approach to tackling its urban development and to ensuring energy efficiency. In August 2011, the city launched its Growth Development Strategy to try to overcome problems of sprawling and population increase. This should be advocated in fast growing cities and emerging new ones in a continuously urbanising world.

3.6 Guidelines for Egyptian Sustainable Cities

The city of 10th of Ramadan is an example of new development in Egypt. It was one of the first new cities to be developed outside existing cities in the 1970s and is situated to the northeast of Cairo, some 50 km away. The city was planned according to the principles of the Athens Charter, advocating separation of uses and car dependency. This has contributed to its limited ability to attract population and to sustain growth. Although the city has a strong industrial economy based in its big industrial zone, currently the city's population is only 260,000 (far behind its original target population of 500,000 by 2005) with only 66% occupation and a density of 42.8 inhabitants/ha. Table 3.2 [1,34] illustrates the poor performance

TABLE 3.1

Cities' Energy Efficiency Performance Indicators and Relevant Sustainable Urbanism Principles

Indicator	City	Value	Compactness				Completeness			Connectedness			Quality of Life
			Walkability	Connectivity	Increased Density	Compact Building Design	Mixed Uses	Mixed Housing	Sense of Place	Integrating Transportation and Land Use	Green Transportation	Variety of Transportation Choices	Preserving Open and Natural Areas
Sustainable urbanism principles	Copenhagen		■	■			■	■		■	■	■	■
	Curitiba		■	■				■		■	■	■	■
	Johannesburg		■	■			■	■		■	■	■	■
Electricity consumption	Copenhagen	80.63 gigajoules/inhabitant											
	Curitiba	743 megajoules of electricity per US$1,000 GDP				—							
	Johannesburg	5.6 gigajoule/inhabitant				—							
Clean and efficient energy policies	Copenhagen	18.76% renewables											
	Curitiba	84% of electricity from hydropower									■	■	
	Johannesburg	Promotes solar power, convert landfill gas into electricity									■	■	

Climate change action plan	Copenhagen	Campaigns to motivate lifestyle change and citizens' involvement in developing solutions	
	Curitiba	Monitoring CO_2 absorption rate in green spaces, and total city CO_2 emissions	–
	Johannesburg	N/A*	–
Eco buildings policy	Copenhagen	554 mega joules/m^2; two urban development projects to create 'carbon-neutral neighbourhoods', with low-energy buildings, sustainable energy networks and environmentally friendly transport	
	Curitiba	No eco-efficient standards for buildings, no incentives to lower energy use	–
	Johannesburg	N/A	–
Green spaces per capita	Copenhagen	80% live within 300 m of a park or recreation area	
	Curitiba	52 m^2/person	
	Johannesburg	230.7 m^2/person	–
Population density	Copenhagen	105 person/ha	
	Curitiba	43 person/ha	
	Johannesburg	24 person/ha	
Land use policy	Copenhagen	Redevelopment of brown field sites	
	Curitiba	Incentivise landowners to establish public parks on their private land and control sprawl	
	Johannesburg	Reduce urban sprawl by rehabilitating underpopulated centre neighbourhoods and building new mixed-density and mixed-income housing developments with access to services and transport	

Continued

TABLE 3.1 (*Continued*)

Cities' Energy Efficiency Performance Indicators and Relevant Sustainable Urbanism Principles

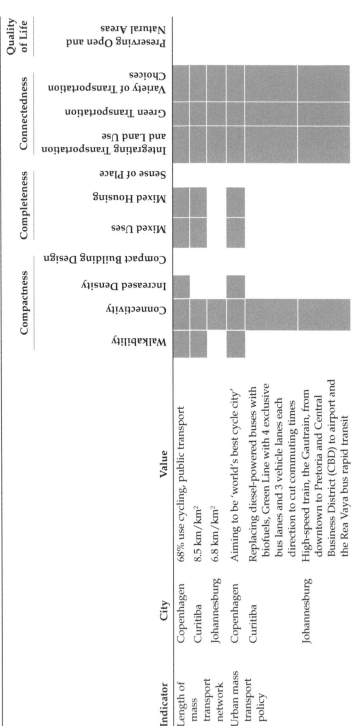

Indicator	City	Value
Length of mass transport network	Copenhagen	68% use cycling, public transport
	Curitiba	8.5 km/km²
	Johannesburg	6.8 km/km²
Urban mass transport policy	Copenhagen	Aiming to be 'world's best cycle city'
	Curitiba	Replacing diesel-powered buses with biofuels, Green Line with 4 exclusive bus lanes and 3 vehicle lanes each direction to cut commuting times
	Johannesburg	High-speed train, the Gautrain, from downtown to Pretoria and Central Business District (CBD) to airport and the Rea Vaya bus rapid transit

Congestion reduction policy	Copenhagen	Increase % of regularly using a bicycle to get to work or education from 36% to 50% by 2015 to reduce congestion
	Curitiba	Traffic light sequencing, traffic information systems, dedicated delivery times and access points
	Johannesburg	Solar power signals at key intersections not susceptible to power failures causing congestion and higher fuel consumption from queuing drivers

Source: The energy efficiency performance indicators are derived from: The Economist Intelligence Unit, *European Green City Index: Assessing the Environmental Impact of Europe's Major Cities*, Siemens AG, Munich, 2009; The Economist Intelligence Unit, *Latin American Green City Index: Assessing the Environmental Performance of Latin America's Major Cities*, Siemens AG, Munich, 2010; The Economist Intelligence Unit, *African Green City Index: Assessing the Environmental Performance of Africa's Major Cities*, Siemens AG, Munich, 2011. The related sustainable urbanism principles are according to: Khalil, H., Sustainable urbanism: Theories and green rating systems, 48th AIAA/ASME/SAE/ASEE Joint Propulsion Conference & Exhibit and 10th Annual International Energy Conversion Engineering Conference, 30 July–1 August, 2012, Atlanta, GA.

Note: Adopted principles are highlighted.

* Not applicable.

TABLE 3.2

10th of Ramadan City Energy Efficiency Performance and Adoption of Sustainable Urbanism Principles

Indicator	Compactness					Completeness		Connectedness			Quality of Life
	Walkability	Connectivity	Increase Density	Compact Building Design	Mixed Uses	Mixed Housing	Sense of Place	Integrating Transportation and Land Use	Green Transportation	Variety of Transportation Choices (three wheeler)	Preserving Open and Natural Areas
Sustainable urbanism principles	No pedestrian network, services are too far	Good road network	Density is low and occupancy too	Buildings are mainly compact	Separation of uses	A variety of types and sizes	A sense of incompleteness	Transportation is not well integrated	Not present	Buses, microbuses and auto rickshaw	No maintenance

Sustainable urbanism principles	
Electricity consumption	414.3 megajoules/inhabitant while the world average rate is 684 megajoule/inhabitant; however it is less than the average rate of Egypt is 450 megajoules/inhabitant
Clean and efficient energy policies	No renewable energy or liquefied gas
Climate change action plan	No concern for climate change or responsiveness to city location in an arid zone
Ecobuildings policy	No concern with ecobuildings or improving environmental performance
Green spaces per capita	12.28 m²/person; however, most open spaces are not planted and existing green spaces lack maintenance—in addition to 94 m²/person of forests and agricultural land surrounding the city
Population density	42.8 person/ha but planned to increase mixed-use centres to increase density
Land use policy	Originally planned with low net densities then increased in following districts
Length of mass transport network	No available data
Urban mass transport policy	New metro line is planned to link the city to Cairo
Congestion reduction policy	150,000 workers commuting daily to surrounding towns and villages, a new electric train is planned to reduce congestion from commuting factory buses

Source: The energy efficiency performance indicators are derived from: Associated Consultants and Little, A.D., *10th of Ramadan Strategic Plan*, General Organisation of Physical Planning and Ministry of Housing, Utilities and Urban Development, Cairo, Egypt. 2010. The related adopted sustainable urbanism principles are according to Khalil, H., Sustainable urbanism: Theories and green rating systems, 48th AIAA/ASME/SAE/ASEE Joint Propulsion Conference & Exhibit and 10th Annual International Energy Conversion Engineering Conference, 30 July–1 August, 2012, Atlanta, GA, paper No. AIAA 2012-4248.

Note: Adopted principles are highlighted.

of the city with respect to energy efficiency in indicators and adoption of sustainable urbanism principles.

In order for the city of 10th of Ramadan to improve its performance, it should adopt the principles of sustainable urbanism in its urban development plans. The latest strategic urban development plan was concluded in 2010; however, it is not operational yet. This plan focused on economic development and on increasing the city's efficiency; however, little concern was directed towards more green policies. Thus, the following recommendations can be concluded based on analysis of best practices worldwide:

1. It is vital to increase occupancy by advocating mixed uses to develop currently vacant neighbourhood service centre and to ensure availability of daily needs within walking distances.

2. Increasing densities in existing parts and new districts to enhance walkability and decrease energy consumption.

3. Enhancing cyclability through dedicating separate lanes, especially since the city now is underpopulated (a factor that facilitates implementing such ideas).

4. Raising awareness about climate change plans for municipalities and city residents.

5. Encouraging factory owners to invest in public housing for their workers instead of expanding their bus fleet for daily transporting workers to surrounding towns and villages.

6. Enforcing the current Egyptian energy code for buildings to ensure better performance, especially in commercial and administrative buildings.

7. Developing a comprehensive transportation network advocating fast transit systems to decrease energy consumption and congestion.

8. Speeding up the implementation of the designated electric train to link the city to Cairo.

9. Encouraging the use of solar energy in many applications (such as water heaters or street lighting), especially because the city is in the middle of the desert.

10. Promoting planting designated green areas but using ecological landscapes and choosing suitable plants that are responsive to the arid climate and limited availability of water for irrigation and that require less maintenance.

11. Encouraging recycling to preserve resources and to maximise the benefit from latent energy used during initial manufacturing.

3.7 Conclusions

Energy efficiency is currently the main concern for environmental performance. It is a common denominator in many sectors related to the city: land use management, building construction, transportation, infrastructure and more. This entails a holistic approach to urban management in order to promote green cities. This chapter has shown the strong interrelation between adopting sustainable urbanism principles and achieving high environmental performance. It takes years to change cities that were planned to be car dependent, where uses were separated. However, such continuous effort is crucial because the world is rapidly urbanising; the number of city dwellers has exceeded that of rural inhabitants. There are great opportunities to learn from best practices in various areas worldwide to enhance energy efficiency. However, user commitment remains essential to reverse current consuming behaviour. Moreover, municipalities' adherence to a comprehensive, yet flexible, plan advocating energy efficiency over time, even with a change of personnel, can create more sustainable cites.

References

1. Khalil, H. (2012). Sustainable urbanism: Theories and green rating systems, *48th AIAA/ASME/SAE/ASEE Joint Propulsion Conference & Exhibit and 10th Annual International Energy Conversion Engineering Conference*, 30 July–1 August, 2012, Atlanta, GA, paper No. AIAA 2012–4248.
2. Costanza, R., Fisher, B., Ali, S., Beer, C., Bond, L., Boumans, R., Danigelis, N. L., et al. (2008). An integrative approach to quality of life measurement, research and policy, *Surveys and Perspectives Integrating Environment and Society*, Vol. 1, pp. 11–15.
3. Mercer. (2011). *Mercer 2011 Quality of Living Survey Highlights—Defining 'Quality of Living'*, Mercer, Retrieved January 25, 2012, from http://www.mercer.com/articles/quality-of-living-definition-1436405
4. The Economist Intelligence Unit (EIU). (2007). The Economist Intelligence Unit's quality-of-life index, the world in 2005, *The Economist*, Retrieved January 20, 2012, from http://www.economist.com/media/pdf/QUALITY_OF_LIFE.pdf
5. Nation Ranking. (2011). *Quality of Life Index 2011 Rankings*, Nation Ranking: Quantifying the World of Sovereign States, Retrieved January 21, 2012, from https://nationranking.wordpress.com/category/quality-of-life-index/
6. Organization for Economic Cooperation and Development (OECD). (2011). *Better Life Initiative Executive Summary*, OECD Better Life Initiative, Retrieved January 20, 2012, from http://oecdbetterlifeindex.org/
7. Newman, P. (2006). The environmental impact of cities, *Environment and Urbanization*, Vol. 18, No. 2, pp. 275–295.

8. Devuyst, D., Hens, L., and De Lannoy, W. (2003). *How Green Is the City? Sustainability Assessment and the Management of Urban Environments*, Columbia University Press, New York.
9. Moos, M., Whitfield, J., Johnson, L. C., and Andrey, J. (2006). Does design matter? The ecological footprint as a planning tool at the local level, *Journal of Urban Design*, Vol. 11, No. 2, pp. 195–224.
10. McHarg, I. (1969). *Design with Nature*, The Natural History Press, Garden City, NY.
11. Hough, M. (1995). *Cities and Natural Process*, Routledge, London.
12. Dramstad, W. E., Olson, J., and Forman, R. (1996). *Landscape Ecology Principles in Landscape Architecture and Land Use Planning*, Island Press, Washington, DC.
13. Gordon, D. and Tamminga, K. (2002). Large-scale traditional neighbour-hood development and pre-emptive ecosystem planning: The Markham experience, 1989–2001, *Journal of Urban Design*, Vol. 7, No. 3, pp. 321–340.
14. Spangenberg, J. H. and Lorek, S. (2002). Environmentally sustainable house-hold consumption: From aggregate environmental pressures to priority fields of action, *Ecological Economics*, Vol. 43, pp. 127–140.
15. Leite, C. and Tello, R. (2011). *Megacity Sustainability Indicators*, Retrieved April 15, 2012, from http://www.stuchileite.com
16. Forum for The Future and General Electric GE. (2010). *The Sustainability Cities Index 2010*, Forum for The Future and GE, London.
17. The Academy of Urbanism. (2010). *The Freiburg Charter of Sustainable Urbanism: Learning from Place*, The Academy of Urbanism, London.
18. Committee for The Development of Environmental Performance Assessment Tool for Cities and Institute for Building Environment and Energy Conser-vation. (2011). *Overview of CASBEE for Cities*, Japan Sustainable Building Consortium (JSBC), Tokyo.
19. Edahiro, J. (2009). *Conceptual Basis of the Movement to Create and Propagate 'Eco-Model Cities' Initiatives of the Japanese Government*, JFS Newsletter No. 78, Retrieved June 8, 2011, from http://www.japanfs.org/en/mailmagazine/newsletter/pages/028824.html
20. Murakami, S. (2008). Promoting eco-model cities to create a low-carbon society, *International Seminar on Promoting the Eco-Model Cities for the Low Carbon Society*, Tokyo, Japan.
21. Farr, D. (2008). *Sustainable Urbanism: Urban Design with Nature*, John Wiley, Featherstone, NJ.
22. Congress for the New Urbanism, Natural Resources Defense Council, and U.S. Green Building Council. (2012). *LEED 2009 for Neighborhood Development Rating System*, U.S. Green Building Council, Washington, DC.
23. The Economist Intelligence Unit. (2009). *European Green City Index: Assessing The Environmental Impact of Europe's Major Cities*, Siemens AG, Munich.
24. *Green City Index*. (2012). Siemens, Retrieved May 7, 2012, from http://www.siemens.com/entry/cc/en/greencityindex.htm
25. Holden, E. and Norland, I. T. (2005). Three challenges for the compact city as a sustainable urban form: Household consumption of energy and transport in eight residential areas in the greater Oslo region, *Urban Studies*, Vol. 42, No. 12, pp. 2145–2166.
26. NÆSS, P. (1997). *Fysisk Planlegging Og Energibruk* [Physical Planning and Energy Use], Tano Aschehoug, Oslo.

27. Tirado, J. (n.d.). *Smart Growth Advances Nationally*, New Urbanism, Retrieved December 24, 2009, from www.newurbanism.org

28. Khalil, H. (2009). Energy efficiency strategies in urban planning of cites, *Seventh International Energy Conversion Engineering Conference IECEC 2009*, AIAA, Orlando.

29. *Principles of New Urbanism*. (n.d.). New Urbanism, Retrieved December 24, 2009, from http://www.newurbanism.org/newurbanism

30. Khalil, H. (2010). New urbanism, smart growth and informal areas: A quest for sustainability, In Steffen Lehmann, Husam AlWaer and Jamal Al-Qawasmi (eds.), *Sustainable Architecture & Urban Development*, CSAAR, Amman, pp. 137–156.

31. Transit Oriented Development. (n.d.). *Design for a Livable Sustainable Future*, Transit Oriented Development, Retrieved January 29, 2010, from http://www.transitorienteddevelopment.org

32. The Economist Intelligence Unit. (2010). *Latin American Green City Index: Assessing the Environmental Performance of Latin America's Major Cities*, Siemens AG, Munich.

33. The Economist Intelligence Unit. (2011). *African Green City Index: Assessing the Environmental Performance of Africa's Major Cities*, Siemens AG, Munich.

34. Associated Consultants and Little, A. D. (2010). *10th of Ramadan Strategic Plan*, General Organisation of Physical Planning and Ministry of Housing, Utilities and Urban Development, Cairo.

4

Energy-Efficient Informalisation*

4.1 Introduction: Scarce Resources, Efficient Practices, Urbanisation and Sustainability

Urbanisation is the leading sector on which humans are working; the ever-growing population is spreading globally and constantly moving into cities. Major urbanisation activities are taking place in the developing world. Over the past 50 years, cities have expanded into the land around them at a rapid rate. Highways and transport systems have been built, and valuable farmland has been lost. Globally, urbanisation levels will rise dramatically in the next 40 years—by 70%, with a population of almost 6.4 billion (UN Population Division 2007). According to the UN-Habitat 2008 report 'The State of the World's Cities 2008/2009: Harmonious Cities', every day 193,107 new city dwellers are added to the world's urban population, which translates to slightly more than two people every second. For the years 2000–2005, in developed nations, the total increase in urban population per month was 500,000, with an annual growth rate of 0.54%, compared to 5 million in the developing world, with an annual growth rate of 2.67% [2].

Cities and urban settlements must be prepared to meet this challenge. In order to avoid being victims of their own success, cities must search for ways in which to develop sustainably. A successful city must balance social, economic and environmental needs; it has to respond to pressure from all sides.

There are a number of theories that define sustainable urban development. In the Western world, there are urban movements that combat suburbanisation, such as 'new urbanism', smart growth, Transit Oriented Development (TOD) and sustainable urbanism. They attempt to achieve more environmentally sustainable urbanism. This urbanism, as specified by the UN-Habitat (2009), requires that (a) greenhouse gas emissions are reduced and serious climate change mitigation, and adaptation actions are

---

* Most of this chapter is derived from Khalil [1].

implemented; (b) urban sprawl is minimised, and more compact towns and cities served by public transport are developed; (c) non-renewable resources are sensibly used and conserved; (d) renewable resources are not depleted; (e) the energy used and the waste produced per unit of output or consumption is reduced; (f) the waste produced is recycled or disposed of in ways that do not damage the wider environment; and (g) the ecological footprint of towns and cities is reduced [3].

Although cities in the developed world grow in a planned pattern, following one code or the other, compact or dispersed, those in the developing world mostly grow informally. This informal growth poses an additional challenge to the urbanisation process itself. The UN-Habitat agenda stresses that cities, in their quest for development, should consider social aspects. These aspects are (a) equal access to and fair and equitable provision of services; (b) social integration by prohibiting discrimination and offering opportunities and physical space to encourage positive interaction; (c) gender and disability sensitive planning and management; and (d) the prevention, reduction and elimination of violence and crime [3].

This chapter addresses the issue of informal urbanisation in the developing world and compares it to sustainability criteria as specified by the new trends of sustainable urbanism. The chapter highlights the green aspects of this form of urbanisation in the developing world, with a special reference to Egypt, as opposed to the dominant idea of unsafe, informal and illegal areas. These sustainability potentials can be the groundwork for upgrading programmes in addition to the provision of shelter, water, sanitation and services.

4.2 Western Urbanisation and the Call for Sustainable Urban Development

Urban settlements in the developed world are expanding mainly by suburbanisation and the creation of new settlements. There is much debate about how to urbanise sustainably—how to combat the disadvantages of current planning theories ruling urban expansion. There are mainly two trends or poles driving urban planning and growth—compact and dispersed development. Currently there are several theories and practices that are leading urban design and planning in the Western hemisphere; these are new urbanism, smart growth, transit-oriented development and sustainable urbanism. They all share common aspects and principles (as discussed in Chapter 3). They all aspire to achieve better energy efficiency and to provide improved quality of life for city dwellers. This chapter will discuss the relationship between these aspects of sustainable urban design and planning and informal areas in the developing countries.

4.3 The Developing World Urbanisation: Informalisation

Informal growth is the dominating pattern in the urbanisation process all over the developing world. It is how people meet their needs when their governments fail to do so. Informalisation can be defined as 'a process which is unregulated by the institutions of society in a legal and social environment in which similar activities are regulated' [4].

Roy states that there are two different perspectives dominating debates about informality. The first one is derived from the 'Urban 21 Report' published in *Urban Future 21: A Global Agenda for 21st Century* by Sir Peter Hall and Ulrich Pfeiffer. They see informally growing cities as deteriorating, decaying and uncontrolled. The second perspective is pioneered by Hernando De Soto in *The Mystery of Capital* (2000), where he considers informality as a heroic adventure. He views an informal economy as people's natural and creative reaction to government inefficiency in supplying basic needs for the poor. However, these two perspectives represent two antidotes, where neither is solely true in itself. Instead, informality can be seen as a pattern for urbanisation instead of being an adversary of the formal sector, providing a promising resource instead of a catastrophe [5].

A recent definition of informal areas in Egypt is 'all what is self-built, whether single or multi storey buildings or shacks, in the absence of law and urban regulations enforcement. They are areas built on land not allocated for construction as specified in the city urban plan. Despite the fact that the buildings' conditions may be good, they might be unsafe environmentally and socially, and or lacking basic infrastructure and services' [6].

There are mainly two types of informal areas: squatter areas and informal subdivisions. Squatter areas are mainly chaotic, unplanned and marginal; informal subdivisions are subdivided land with legal ownership, but that lack infrastructure and areas for public services and uses [7].

In Egypt, the general classification of informal areas includes the informal areas built on agricultural land and areas built on desert land (the two main patterns of informal areas) in addition to shacks and environmentally unsafe areas.

4.3.1 Informal Areas Built on Agricultural Land

These areas are illegal because they are built on agricultural land not allocated for construction, and they defy the banning of mixed uses as specified by law. Their general characteristics are: (a) narrow long streets with a width of no more than 4–5 meters, some even with dead ends; (b) regular block shapes according to agricultural basins subdivisions; and (c) housing units having constant depth but with different street frontage. Heights are according to owner's affordability.

4.3.2 Informal Areas Built on Desert Land

These appear on vacant land on city fringes where the land is publicly owned, making the buildings there illegal. Their general characteristics are: (a) curved, uneven streets; (b) temporary houses made of primitive materials such as tin, carton or straw; and (c) insecure tenure.

4.3.3 Informal Area Upgrading Programmes

Informal areas and slums pose a significant threat to the green agenda because many are built on physically unsafe land that is vulnerable to natural hazards. They often deprive the city of surrounding agricultural land and foreshore land for flood control and natural biofiltration from fringing wetland vegetation. Severe erosion can result from steep slopes when they are settled upon.

Informal area and slum upgrading programmes are mainly concerned with a totally different agenda. These programmes are oriented towards the brown agenda, with five main dimensions: access to safe water, sanitation, providing secure tenure, durable housing, and sufficient living area. These are the indicators set by UN-Habitat to identify the improvement of slum dwellers as stated by Target 11 of the Millennium Development Goals (MDGs). However, this focus neglects other aspects of environmental sustainability such as reducing greenhouse gas emissions, minimising sprawling, sensibly using non-renewable resources, waste recycling, reducing energy consumption and reducing cities' ecological footprints.

4.4 Do Informal Areas Possess Sustainable Potentials? Is Informalisation 'Smart'?

Usually, when informal areas or slums are addressed, the brown agenda and its relative indicators are the main concern. However, these areas are rarely seen as having green aspects or sustainable characteristics. Citing four examples, the following section discusses the presence of many of the sustainability aspects as stated in new urban planning trends for informal areas in the Egyptian context. Boulaq Al-Dakrour and Giza's northern sector (Imbaba) are two informal districts of Cairo, each with around 700,000–900,000 inhabitants. The other examples are two upgrading projects, one located in vacant land at the relocated Imbaba airport and the other in Zeinhom district, a formerly shack area that was demolished.

4.4.1 Defined Neighbourhood (with Quality Architecture and Urban Design)

Informal areas usually have defined edges that separate them from their surrounding areas—such as a railroad, canal, or ring road. They have a distinct urban pattern, especially those following agricultural basin subdivisions. However, they do not have centres; instead, the main streets act as the centres with a concentration of uses and markets. Secondary streets act as recreational spaces where children play due to the prevailing sense of security (see Figure 4.1) [8].

What these areas lack is an overall urban vision because they are built incrementally. Moreover, where the land is privately owned, open spaces are usually overlooked because they have little or no economic value. Also, when there is an open space, it is not taken care of unless there is a strong sense of community among residents. As for the quality of architecture, in some areas, it can be poor but can be upgraded to promote local character and sense of place.

However, the prevailing visual image is homogeneous due to building with the same materials, bricks and concrete, and following the same urban pattern (see Figure 4.2).

4.4.2 Compactness

Informal areas are compact in density (e.g.: 890 person/ha in Boulaq Al-Dakrour district), exceeding other formal areas due to the private development mechanism, thereby providing a perfect setting for walkability and energy efficiency. The buildings are stacked together with usually only one free facade that minimises thermal loads, maximises space use and enhances energy efficiency. Compactness can be explained through the following aspects.

4.4.2.1 Walkability

Informal areas are characterised by narrow streets that are mainly pedestrian. Services, which are mainly community built, are usually within less than a 10-minute walk (see Figure 4.3). However, government provided services may not exist in close proximity.

4.4.2.2 Connectivity

Streets are interconnected; however, they are more favourable for pedestrians than vehicles because they are narrow. Although there is a network of wider streets, due to increased traffic load, they are usually congested—especially at marketplaces and area entrance points. Moreover, streets can be too long and without crossings, which decreases connectivity (see Figure 4.4) [9].

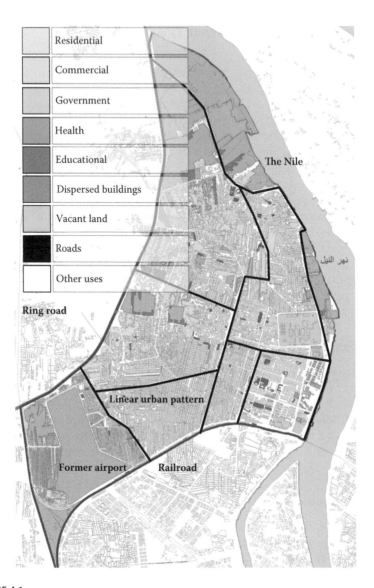

FIGURE 4.1
Giza northern sector land uses. (From Ministry of Housing, Utilities and Urban Development, *Project for Giza Northern Sector Development and Reuse of Imbaba Airport Site*, Ministry of Housing, Utilities and Urban Development, Cairo, 2008.)

4.4.2.3 Increased Density

Informal areas are characterised by very high densities (e.g.: in Boulaq Al-Dakrour, this ranges from 490 to 2,500 persons/ha); usually upgrading programmes tend to relocate some families to decrease this density, if possible. However, the high density should be maintained, unless it poses a threat

(a)   (b)

FIGURE 4.2
Homogeneous buildings in Boulaq Al-Dakrour district, Cairo. (a) Buildings on AlAmer street
as one of the typical mid sized streets in the area, and (b) Buildings on Zenin Street as one of
the main streets in the area.

FIGURE 4.3
Community hospital, nursery and vocational centre located within 300–500 m for daily
accessibility.

to human living conditions. The UN-Habitat suggests a maximum crowding
indicator of two persons/room.

4.4.3 Completeness with Daily and Lifelong Utilities

These areas have a variety of uses that make them complete and independent
for daily needs. They can be considered as a separate identity and provide
lifelong utilities for many residents. Residents can spend their life working,

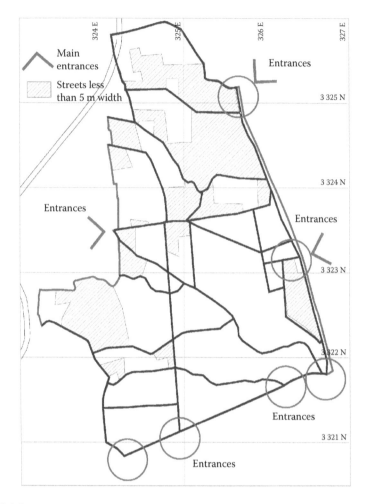

FIGURE 4.4
Street hierarchy and entrances in Boulaq Al-Dakrour, Cairo. (From General Organisation of Physical Planning, Giza Governorate, GTZ, and Ministry of Housing, Utilities and Urban Development, *Boulaq Al-Dakrour District Guide Plan 2004–2017*, GTZ, Cairo, 2004.)

living in an informal area without having to go outside except for some higher educational and health services. This can be apparent through the following aspects.

4.4.3.1 Mixed Use and Diversity

Informal areas are usually characterised by mixed uses that are seen by the formal authorities as incorrect. However, it is the mixed uses that give the area its richness, liveliness and advantage of availability of needs within

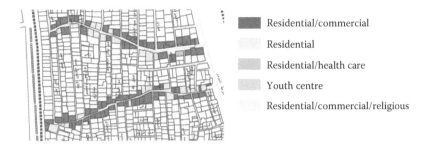

FIGURE 4.5
Mixed uses prevail in streets of Boulak Al-Dakrour, Cairo. (From GTZ, *Participatory Development in Urban Areas Program, Boulaq Al-Dakrour District*, GTZ, Cairo, 2007.)

the area—an advantage sought by the new urban trend plans (Figure 4.5). These areas also offer uses and utilities for a diverse spectrum of groups, ages and incomes [10].

4.4.3.2 Mixed Housing

There is a variety of housing opportunities in informal areas because they are community built and are driven by community needs. Different sizes are available, and in some areas there is a variety of standards, especially in informal areas built on private lands. In Boulak Al-Dakrour, Giza, Egypt, there are buildings with high standard apartments and other buildings accommodating more than one household per unit. This diversity and community-driven development pattern adds to the area's sense of place, as opposed to the identical blocks in publicly developed projects.

4.4.4 Connectedness with Integrating Transportation and Land Use

Informal areas have their own transportation network and modes that might be exclusive to them, which is why entrances to these areas are always considered as transportation nodes. They rely on private transportation modes, ranging from microbuses, pickup trucks and recently auto rickshaws (toktok). These are usually in bad condition and driven by unlicensed drivers, thus posing a threat to passengers' lives and to the environment. Sometimes these areas are built beside railways, underground metro or transportation stations and can benefit from such facilities. Despite this drawback in transportation in informal areas, people use bicycles a lot or they walk, thus adding to the area's green advantages. Land uses are distributed according to needs; commercial uses and other services are distributed along the main streets, leaving the narrower streets for residential and recreational uses.

4.4.5 Enhancing the Quality of Life

Because informal areas usually grow on undeveloped land; whether it is agricultural or a high risk zone, these areas do not provide access to nature. On the contrary, they are a threat to nature. This can be overcome by preserving whatever vacant land still exists inside the informal areas and by belting the whole area to prevent its further encroachment on green fields. However, because these areas are usually on the fringes of cities, they have the advantage of proximity to nearby nature. The quality of life of informal area residents can be enhanced through addressing the attributes of the brown agenda and by emphasising green advantages, providing an outlet for overcrowded residents to enjoy nature.

4.4.6 Stakeholder Participation

Informal areas are self-built by the community and its informal sector. Almost all development decisions are community directed. The community has succeeded in providing housing and basic services through collaborative efforts. Other services, such as child care facilities, medical centres, training centres and so forth, are usually provided through community-based organisations. Thus, these efforts provide a solid ground for further participation.

Most upgrading programmes focus on the five main dimensions in accordance with the UN-Habitat indicators, namely: access to safe water, sanitation, providing secure tenure, durable housing and sufficient living area. In this quest, these programmes might overlook the green potentials of the informal area to the extent that they might decrease them. Rebuilding projects provide a perfect opportunity to build on the green potential of informal areas and to avoid their pitfalls.

In the government-managed Imbaba housing project, a community piazza is designed to be the heart of the neighbourhood (13,896 inhabitants on 22 hectares). The area is designed with two main mixed use axes that link the areas to the adjacent park and lead to the main piazza with its community facilities (e.g.: nursery and mosque). The design promotes pedestrianisation with proximity of services within 600 m. The density is 630 persons/ha, resembling the surrounding district, and commercial uses are integrated with residential buildings. The area provides a diversity of housing alternatives with one-, two-, and three-bedroom flats and handicapped adapted units, thus adding completeness to the area. However, no green transportation was integrated into the project, only linking the area to the public bus route and to the public/private microbus service. The project aspires to enhance the quality of life of families relocated due to the demolishing of their former houses during main street widening in the comprehensive plan to upgrade Giza's northern sector.

Moreover, the nearby new park is one of the ambitious projects to improve the quality of life for the whole sector. This project has special significance because there have been a lot of debates about it in the media in the years since the relocation of Imbaba airport and the availability of its former site for development. Various stakeholders participated in different stages—from the preliminary concept to the various surveys to assessing needs and priorities (Figure 4.6) [8].

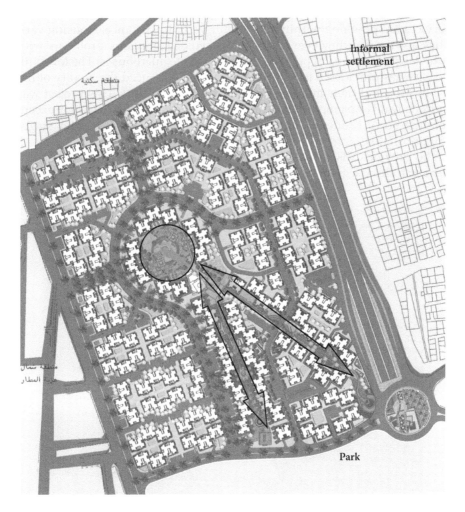

FIGURE 4.6
Imbaba Airport housing area, Cairo. (From Ministry of Housing, Utilities and Urban Development, *Project for Giza Northern Sector Development and Reuse of Imbaba Airport Site*, Ministry of Housing, Utilities and Urban Development, Cairo, 2008.)

The Zeinhom Gardens housing project was a different case. The site was formerly a shack area and was designated for demolishing. The project, initiated by the government and the Red Crescent NGO, was implemented on three stages. The first and second stages had typical designs for buildings with limited variations in colours; however, the third stage was adapted to the contextual Islamic heritage of old Cairo. The area is designed with pedestrian-friendly streets and paths; however, densities were lowered in the first stage to 388 persons/ha and then increased in later stages to 580 persons/ha. Commercial facilities were built separately and were not allowed to be integrated into residential buildings; however, services are provided within a 10-minute walking distance. Housing varieties are limited; however, each phase avoided previous problems with unit design. The area depends on public/private microbuses. The low built density has provided a lot of open spaces that enhanced the quality of life in the area, but their maintenance requires much effort. This project was designed and implemented with minimum community participation except for first stage residents' requests that were integrated into later stages (Figure 4.7) [11].

The previous discussion can be clearly seen through Table 4.1, which demonstrates the green aspects of examples of informal areas.

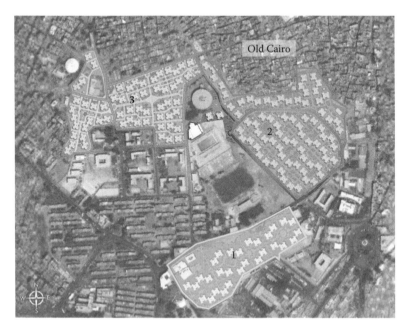

FIGURE 4.7
Zeinhom Gardens housing project with its three stages, Cairo. (From Cairo Governorate and Housing and Utilities Sector, *Zeinhom Gardens Housing Project*, Cairo Governorate, Cairo, 2009.)

TABLE 4.1

Analysis of Sustainability Aspects of Existing and Rebuilt Informal Areas

Sustainable Urban Development	Giza's Northern Sector	Boulaq Al-Dakrour District	Zeinhom Gardens Housing Project	Imbaba Airport Housing Project
Defined neighbourhood: quality architecture and urban design	Defined boundaries and sense of security	Defined boundaries and sense of security and homogeneous buildings	Third phase adapting to contextual Islamic heritage	Community piazza with mixed uses to promote sense of place and urban life
Compactness: walkability, connectivity, and increased density	Pedestrian-friendly streets; high densities with up to 890 persons/ha	Pedestrian-friendly streets; high densities with up to 650 persons/ha	Pedestrian-friendly streets; density of 388 persons/ha then amended in third phase	Promoting pedestrianisation; new area with density of 630 persons/ha
Completeness: mixed use and diversity, mixed housing	Uses are mixed; housing is needs-driven	Uses are mixed; housing is needs-driven	Separate area for commercial market; identical units, but each phase avoided previous problems	Commercial use is integrated into residential buildings 1-, 2-, and 3- bedroom apartments with handicapped adapted units
Connectedness: green transportation	Privately owned and operated with an underground metro under construction	Privately owned and operated with underground metro on the edge	Private and public mode of transportation	Private and public mode of transportation
Enhancing quality of life	Strategic planning with parks and open spaces	Local upgrading programmes under implementation	Rebuilding with better quality and open spaces	Area is linked to adjacent park
Stakeholder participation in development decisions	Initially developed by community	Initially developed by community	Minimal participation in the design process	Stakeholder participation in assessing needs and priorities

4.5 Conclusions

Unmanaged or chaotic urban growth is considered a significant obstacle to sustainable urban development. In a report about Egypt's achievement of the MDGs, it was recommended to plan and develop environmentally sustainable urban communities with affordable low-cost housing and improved operation and maintenance of water and sanitation services [12].

Because more than 60% of Egypt's urban development is informal, it is vital to study informal areas and to analyse their sustainability. Through case study analysis, some of their green characteristics are highlighted. They are compact, walkable and dense. They provide for lifelong needs with a variety of housing and mixed uses. Initially, they are needs driven and depend mainly on their communities' participation in development. However, other absent sustainability aspects should be introduced such as green transportation and promoting access to nature. This can be done by controlling the encroachment of informal areas on agricultural land (belting informal areas) and by providing more sustainable housing opportunities and transit-oriented development. However, the case might be different in slums and in squatter areas because they tend to be more environmentally hazardous.

The people's way of developing is 'smart'; however, their settlements need a more comprehensive approach to ensure the provision of needs, housing and services in a more environmentally responsive pattern. This must be the basis for projects—whether upgrading or redevelopment. The redevelopment projects show that it is important to maintain high densities, walkability, mixed uses and mixed housing and to integrate with nearby open spaces. Providing access to safe water, sanitation, secure tenure, durable housing and sufficient living area should not divert projects from attending to other sustainability aspects.

On the other hand, new formal developments—especially for medium and upper strata—should not overlook the lessons learned from informal development. Low-density dispersed suburbanisation is not the way for environmentally sustainable development. Henceforth, there is a need to develop theories and guidelines for formal planning to address sustainability issues in both informal areas and their upgrading programmes and new formal development deduced from the liveliness of informal areas and global guidelines for sustainable urban development and more locally tailored theories and practices.

References

1. Khalil, H. (2010). New urbanism, smart growth and informal areas: A quest for sustainability, In Steffen Lehmann, Husam AlWaer and Jamal Al-Qawasmi (eds.), *Sustainable Architecture & Urban Development*, CSAAR, Amman, pp. 137–156.

2. United Nations Human Settlements Programme (UN-Habitat). (2008). *State of the World's Cities 2008/2009: Harmonious Cities*, Earthscan, London, UK.
3. United Nations Human Settlements Programme (UN-Habitat). (2009). *Planning Sustainable Cities: Policy Directions, Global Report On Human Settlements*, Abridged Edition, Earthscan, London, UK.
4. Oldham, L., Shorter, F., and Tekçe, B. (1994). *A Place to Live, Families and Child Health in a Cairo Neighbourhood*, The American University in Cairo Press, Cairo, p. 10.
5. Roy, A. (2005). Urban informality: Toward an epistemology of planning, *Journal of The American Planning Association*, Vol. 71, No. 2, pp. 147–158.
6. General Organization of Physical Planning (GOPP) and United Nations Development Program (UNDP). (2007). *Improving Living Condition within Informal Settlements through Adopting Participatory Planning: General Framework for Upgrading and Controling Informal Areas*, GOPP, Cairo.
7. Imperato, I. and Ruster, J. (2003). *Slum Upgrading and Participation: Lessons from Latin America*, The World Bank, Washington, DC.
8. Ministry of Housing, Utilities and Urban Development. (2008). *Project for Giza Northern Sector Development and Reuse of Imbaba Airport Site*, Ministry of Housing, Utilities and Urban Development, Cairo.
9. General Organisation of Physical Planning, Giza Governorate, GTZ, and Ministry of Housing, Utilities and Urban Development. (2004). *Boulaq Al-Dakrour District Guide Plan 2004–2017*, GTZ, Cairo.
10. GTZ. (2007). *Participatory Development in Urban Areas Program, Boulaq Al-Dakrour District*, GTZ, Cairo.
11. Cairo Governorate and Housing and Utilities Sector. (2009). *Zeinhom Gardens Housing Project*, Cairo Governorate, Cairo.
12. Ministry of Economic Development. (2008). *Egypt Achieving The Millennium Development Goals: A Midpoint Assessment*, Ministry of Economic Development, Cairo, Egypt.

5

Energy Generation Plants and Leakages of Energy in Urban Egypt

5.1 Conventional Power Plants

5.1.1 General

In general, the construction and operation of a power plant requires the existence of some conditions such as water resources and stable soil type. There are other criteria that, although not required for the power plant, should be considered—such as population centres and protected areas—because they will be affected by either the construction or operation of the plant. The following list covers most of the factors that should be studied and considered in the selection of proper sites for power plant construction:

1. Transportation network: Easy access and enough access to a transportation network are required during power plant construction and during operation periods.
2. Gas pipe network: Vicinity to the gas pipes reduces the required expenses.
3. Power transmission network: To transfer the generated electricity to the consumers, the plant should be connected to an electrical transmission system. Therefore, the nearness to the electrical network can play a roll.
4. Geology and soil type: The power plant should be built in an area with soil and rock layers that can stand the weight and vibrations of the power plant.
5. Earthquake and geological faults: Even weak and small earthquakes can damage many parts of a power plant intensively. Therefore, the site should be far enough away from faults and previous earthquake areas.
6. Topography: It has been proven that high elevation has a negative effect on production efficiency of gas turbines. In addition, changing a sloping area into a flat site for the construction of the power plant

increases the cost. Therefore, the parameters of elevation and slope should be considered.

7. Rivers and floodways: Obviously, the power plant should be a reasonable distance away from permanent and seasonal rivers and floodways.

8. Water resources: Different amounts of water are required for the construction and operation of a power plant and could be supplied from either rivers or underground water resources. Therefore, having enough available water in a defined vicinity can be a factor in site selection.

9. Environmental resources: Power plant operation has important impacts on the environment. Therefore, priority should be given to locations that are far enough from national parks, wildlife, protected areas, and so on.

10. Population centres: For the same reasons, the site should be far enough from population centres.

5.1.2 Need for Power

In general, the site should be near areas where there is greater need for generation capacity to decrease the amount of power loss and transmission expenses.

1. Climate: Parameters such as temperature, humidity, wind direction and wind speed affect the productivity of a power plant and always should be taken into account.

2. Land cover: Some land cover types such as forests, orchards, agricultural land and pastures are sensitive to pollution caused by a power plant. The effect of the power plant on the land cover types surrounding it should be considered.

3. Area size: Before any other consideration, the minimum area size required for the construction of the power plant should be defined.

4. Distance from airports: Usually, a power plant has high towers and chimneys and large volumes of gas. Consequently, for security reasons, they should be located away from airports.

5. Archaeological and historical sites: Usually historical buildings are fragile and at the same time very valuable. Therefore, the vibration caused by a power plant can damage them, and a defined distance should be considered.

5.1.3 Characteristics of a Steam Power Plant

The desirable characteristic for a steam power plant are as follows:

1. Higher plant efficiency
2. Lower initial cost

3. Ability to use low-quality, high-ash content and inferior fuels
4. Reduced environmental impact in terms of air pollution
5. Reduced water requirement
6. Higher reliability and availability

5.1.4 Classification of Power Plant Cycle

Power plant cycles are generally divided into the following classical groups [1,2]:

1. Vapour power cycle—such as the Carnot cycle, Rankine cycle (Figure 5.1), regenerative cycle, reheat cycle, binary vapour cycle
2. Gas power cycles—such as the Otto cycle, diesel cycle, dual combustion cycle, gas turbine cycle

5.1.4.1 *Rankine Steam Cycle*

The Rankine cycle is the most common of all power generation cycles and is diagrammatically depicted in Figures 5.2 and 5.3. The Rankine cycle was devised to make use of the characteristics of water as the working fluid. The cycle begins in a boiler (State 4 in Figure 5.3), where the water is heated in a constant pressure process until it reaches saturation. Once saturation is reached, further heat transfer takes place at a constant temperature until the working fluid reaches a quality of 100% (State 1). At that point, the high quality vapour is expanded isentropically through an axially bladed turbine stage to produce shaft work. The steam then exits the turbine (at State 2).

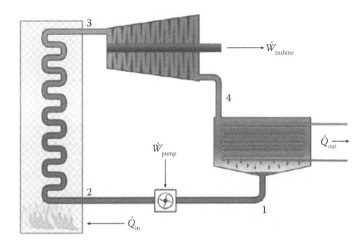

FIGURE 5.1
Simple Rankine power cycle.

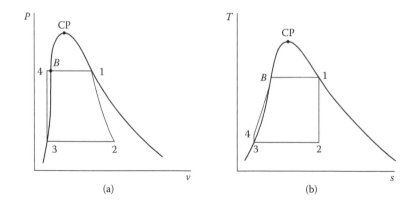

FIGURE 5.2
Diagram for a simple ideal Rankine cycle. (a) on P-V diagram and (b) on T-S diagram.

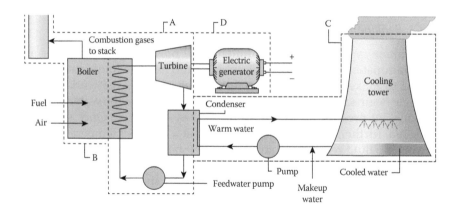

FIGURE 5.3
Schematic of a simple ideal Rankine cycle.

The working fluid (at State 2) is at a low pressure but has a fairly high quality, so it is routed through a condenser, where the steam is condensed into liquid (State 3). Finally, the cycle is completed via the return of the liquid to the boiler, which is normally accomplished by a mechanical pump.

The efficiency of a Rankine cycle is usually limited by the working fluid. Without the pressure becoming supercritical, the temperature range over which the cycle can operate is quite small; turbine entry temperatures are typically around 565°C (the creep limit of stainless steel) and condenser temperatures are around 30°C. This gives a theoretical Carnot efficiency of around 63% compared with an actual efficiency of 42% for a modern coal-fired power station. This low turbine entry temperature (compared with a gas turbine) is why the Rankine cycle is often used as a bottoming cycle in combined cycle gas turbine power stations.

The working fluid in a Rankine cycle follows a closed loop and is reused constantly. The water vapour and entrained droplets often seen billowing from power stations are generated by the cooling systems (not from the closed loop Rankine power cycle) and represent the waste heat that could not be converted to useful work. Note that cooling towers operate using the latent heat of vapourisation of the cooling fluid. The white billowing clouds that form in cooling tower operation are the result of water droplets that are entrained in the cooling tower airflow; they are not, as commonly thought, steam. Although many substances could be used in the Rankine cycle, water is usually the fluid of choice due to its favourable properties, such as non-toxicity and non-reactive chemistry, abundance and low cost, as well as its thermodynamic properties.

5.1.4.2 Gas Cycles

One of the principal advantages the Gas cycles have over other cycles is that during the compression stage relatively little work is required to drive the pump due to the working fluid being in its liquid phase at this point. By condensing the fluid to liquid, the work required by the pump will only consume approximately 1%–3% of the turbine power and so give a much higher efficiency for a real cycle. The benefit of this is lost somewhat due to the lower heat addition temperature. Gas turbines, for instance, have turbine entry temperatures approaching 1500°C. Nonetheless, the efficiencies of steam cycles and gas turbines are fairly well matched.

5.2 Strategic Urban Planning Programme: Lacking Energy Component

Strategic planning is a technique that has been applied to many facets of human activity; the application of strategic planning in urban contexts—in cities, regions and other metropolitan areas—is a relatively recent development. Its beginnings were eminently practical and artistical: a mixture of thought, techniques and art or expertise.

Fifteen years of practice proved to be enough time for the technique to spread and for the first 'Meeting of American and European Cities for the Exchange of Experiences in Strategic Planning' to be organised. Institutions sponsoring the meeting, held in Barcelona in 1993, included the Inter-American Development Bank, the European Community Commission and the Iberoamerican Cooperation Institute. The cities of Amsterdam, Lisbon, Lille, Barcelona, Toronto and Santiago de Chile participated, among others.

At that meeting, it was demonstrated—along with other relevant aspects— that if cooperative processes are used in large cities in order to carry out strategic planning processes, and if a reasonable degree of comprehension is reached between the administration, businesses and an ample representation of social agents, organisational synergies will develop that will eventually improve resource management and citizens' quality of life.

Strategic urban planning (SUP), hence, has dominated the urban planning field for the past two decades or more. It aims to improve the city's performance and to ensure that future development is planned and responsive. Its reliance on a participatory approach enriches the process, promotes local ownership and ensures—to an extent—its actual implementation. The overall aim of the process is to improve the performance of the city through adopting the strategic urban development plan. Its general objectives include clarifying which city model is desired and working toward that goal, coordinating public and private efforts, channelling energy, adapting to new circumstances and improving the living conditions of the citizens affected. Thus, it can be viewed as a tool to enhance the quality of life of city residents.

In the United States, SUP processes, also known as urban renewal projects, began to appear at the end of the 20th century. The city of San Francisco carried out its process between 1982 and 1984. The main motivation behind starting the SUP process was the attempt to adequately react to problematic situations (mainly economic crisis or standstill). At the beginning of the 21st century, this kind of organisation is not reactive but proactive. In the case of Spain, crisis situations are not the main causes of these processes, rather they are motivated by the search for an improved level of public–private cooperation—the wish to coordinate activity, continued improvements, the wish to launch revitalisation processes and even to follow others. The initial determination needed to launch this type of process varies by region; in Spain, most processes (approximately 50%) are fronted by public entities, although a significant percentage has mixed public–private leadership.

5.2.1 Description of SUP Processes

An SUP process is defined as a city project that unifies diagnoses, specifies public and private actions, and establishes a coherent mobilisation framework for the cooperation of urban social actors. A participative process is a priority when defining contents because this process will be the basis for the viability of the objectives and actions proposed. The result of the strategic plan should not necessarily be the creation of regulations or of a government programme (although its adoption by the state and local government should mean the instigation of regulations, investment, administrative measures, policy initiatives etc.) but rather a policy contract between public institutions and civil society. For this reason, the process following the approval of the plan and the monitoring and implementation of measures or actions is just as or more important than the process of elaboration and consensual approval. That is why SUP is now considered a type of governance.

5.2.1.1 Basic Stages of an SUP Process

1. Using the work of the technical secretary as a starting point, work groups debate and approve a diagnosis of the city that includes its localisation. The document must be approved by the executive committee, by the general council, or by a full meeting of the corporation as the case may be.

2. Based on the diagnosis, and keeping in mind its antecedents and conclusions, strengths and weaknesses, the next step is the creation of scenarios and, based on the use of imagination and rigour, the development of prospective tasks related to the creation of future alternatives so that the executive committee can select a model or vision for the city. Their choice will be the basis for the generation of related key topics and/or directions for general actions to be taken.

3. Once the work teams have been reorganised, mainly made up of key decision makers and implementers, each key topic and line of action will be dealt with separately, designing a detailed list of necessary and/or advisable projects. Once the results have been consolidated, a prioritised list of projects will be made available from which a selection will be made. The next step is the elaboration of an action plan that includes the agents involved, timing and resources.

4. Once all of the previously mentioned documents have been approved, the next step is implementation—carrying out the project itself. This stage is decisive; at this point, plans are usually given a structure in which the organisation is even more explicitly clarified.

5.2.1.2 Critical Comments on SUP Processes

Sectors in the area of civic participation, as well as planning professionals and political activists, have all expressed criticism of SUP processes. However, SUP processes include aspects that favour selective participation, territorial organisation and coordination/cooperation between public and private sectors. On the other hand, SUP processes seem to be independent of political ideologies.

5.2.1.3 Theoretical Development

Knowledge relating to SUP processes is evolving in two complementary directions that can be denominated, borrowing concepts from programming, as bottom-up and top-down.

5.2.1.3.1 Bottom-Up

There are clear differences between what could be called the traditional approach to strategic planning and the emerging approach.

1. Before: product predominance; now: process predominance
2. Before: sector specific; now: integrated
3. Before: normative; now: strategic
4. Before: goal oriented; now: cost-benefit oriented
5. Before: urban-offer oriented; now: urban-demand oriented
6. Before: subject to administrative limitations; now: supersedes administrative limitations and enters in metropolitan areas
7. Before: open participation; now: focused participation

Of course, in 2006, there was a clear evolution that attempted to adapt to changes, political sensitivities and even trends. In any case, this is a line of thought and action that takes full advantage of the experience of projects that have already been implemented.

5.2.1.3.2 Top-Down

Given that:

1. The influence of each agent in the global process under consideration for implementation is yet to be determined.
2. There are no generally accepted criteria when creating instruments for measuring progress or regression on the path toward achieving main goals.
3. Cooperation processes among different agents within the city to carry out strategic planning are usually undertaken using a 'framework' organisational structure that highlights differences.
4. Both politics and outside events affect a large city.

This line of research seeks to further the design of a model that will determine the factors related to the success of strategic planning processes in large cities and metropolitan areas. It should be pointed out that a theory explaining SUP in metropolitan areas and/or regions would involve furthering the consolidation of social design.

Since 2008, a national project to prepare strategic urban plans for cities has been launched in Egypt following the issuing of a new building law that mandated a strategic urban plan for each urban agglomeration. This project is administered by the General Organization of Physical Planning (GOPP). In this context, the project for SUP for small cities, funded by the UN-Habitat, aims to improve performance and accountability in planning, implementing, and coordinating action and contributes to achieving the MDGs. The project adopts a decentralised and integrated approach to address three main substantive areas: shelter, basic urban services and local economic development—along with environment, governance and vulnerability as additional crosscutting areas. Because the project design is built mainly on sustainable development, through a participatory process, local stakeholders prepare a strategic urban plan for the following two decades with priority actions to improve housing conditions, urban services and local economy. All gathered data are compiled in urban observatories to contribute to the effective management of urban policy [3].

The process of SUP in all cities in Egypt depends mainly on identifying development projects as the main drivers for achieving future vision and objectives. An initial list of projects is proposed based on feedback from local stakeholders and on current needs identified in the city profile. Stakeholders are divided into four groups representing local administration, local popular council, community-based organisations and the private sector. These four groups are asked to identify priority projects according to their perspectives, and then responses are compiled to consolidate priorities. First priority projects are those that the four groups have identified as priority; second priority projects are those that three groups have identified as priority and so forth for third and fourth priority. The strategic development plan focuses on first and second priority projects as the main drivers for achieving development objectives [4].

It is important to utilise the strategic planning process with its participatory methodology to improve quality of life in cities based on the stakeholders' subjective rating of categories defining their quality of life. However, it is obvious that until now there has been no concern with energy efficiency as an aspect of sustainability. However, this can be correlated with the current low quality of services in Egyptian cities; hence, they are more concerned with their more vital needs.

5.3 Building New Communities in Desert Areas

During the 1970s, Egyptian state envisioned developing new cities to accommodate the ever growing urban population (mainly Cairo) in more appropriate environments in the desert areas surrounding cities or on major development corridors [5]. The first generation included 10th of Ramadan, 15th of May, 6th of October, and Sadat City in the Greater Cairo metropolitan area. Although all new urban communities (NUCs) have been planned as self-sustaining since the late 1970s, they have not yet reached their planned capacities. Moreover, they currently serve either as dorm cities or only workplaces for different groups, increasing daily commuting. This is evident in the city of 10th of Ramadan, a deserted city at night that buzzes during day with its factory workers on their daily trip in from surrounding villages.

Because the main driving force for moving from rural to urban or urban to urban areas is the search for a better job opportunity, it is clear how vital it is to properly plan these new communities. Thus, it is clear that housing should follow jobs and not vice versa. If it is vital to transfer people to improve their quality of living, jobs must be provided first. Otherwise, high-income inhabitants will continue to commute to their original jobs, creating other problems of traffic, transportation and pollution; lower income dwellers would refuse to move because the job–housing package is more essential for them. On the other hand, lack of appropriate housing in the new job location might result in more commuting, raising expenditures or unemployment where people would refuse to move in the search for a better job. In both cases, the situation creates more poverty instead of solving the problem and of course accentuates the issue of mobility and energy for transportation [6]. Until now, NUCs in Egypt have been planned according to the principles of the Athens Charter, advocating separation of uses and car dependency. This has contributed to their limited ability to attract population and to sustain growth.

5.3.1 Strategic Plan Highlights

In order to reserve the current urban trend in new communities, especially in Egypt, and to improve their performance, the principles of sustainable urbanism should be adopted in urban development plans. SUP can be immensely useful. The strategic urban plan should focus on economic development and on increasing the city's efficiency. There should be a focus on promoting green policies. A number of relevant recommended actions, based on analysis of best practices worldwide, would be analysed as follows:

1. Housing: An increasing number of residents in the zone will have a decent, safe place to live, in terms of affordable housing construction programmes, self-help housing projects, migrant farm worker housing construction, new mobile home parks and increased code enforcement for new and existing housing.
2. Economic development: Community and economic development activities and efforts to create employment opportunities and to promote business expansion and entrepreneurship, with micro business loans and technical assistance.
3. Education: Including construction of new school buildings, utilisation of bookmobiles to reach unserved areas, and establishment of special task forces to provide adult and job training, GED achievement and youth employment programmes.
4. Health care services: Would provide mobile medical clinics to travel from school to school and to build permanent medical clinics.

5. Infrastructure: Planned projects would provide clean drinking water, safe sewage disposal systems, established bus services to provide access to jobs and training, and increased fire fighting, paramedic and public safety services.

6. Community facilities: Including construction of three multipurpose community centres and creation of programmes to facilitate volunteer community building activities.

5.3.2 Partnerships

Partnerships include participation by federal, state and local governments, non-profits, area businesses, public schools and the community college.

5.3.3 Leveraging

The Desert Communities Empowerment has identified a range of organisations at the regional, governorate and local level where it intends to seek potential funding and technical assistance. The plan also identifies other sources of funding and support from local communities and organisations in terms of in-kind contributions.

5.3.4 Community Involvement

Outreach to the residents in the designated areas consisted of press releases, articles in local newspapers, surveys, a series of public meetings and creation of special topical subcommittees. The overall plan includes all the local and small communities in the area. There was a concerted effort to include all residents in the area in terms of both low-income and minority status. Efforts were made to present information at meetings and in local newspapers in English and in Spanish, so that residents of the area were fully informed about the planning process, and they were aware that their views were considered to be important in developing the community assessment and the plan. Numerous local organisations; local, county and state agencies; as well as self-help groups provided letters of support for the designation as an empowerment zone.

References

1. Khalil, E. E. (1990). *Power Plant Design*, Gordon and Breech, NY, USA.
2. Khalil, E. E. (2013). *Boiler Furnace Design*, LAP Publishers, Saarbrücken, Germany.

3. General Organization for Physical Planning (GOPP) and United Nations Human Settlements Project (UN-Habitat). (2008). *Terms of Reference for Preparing Strategic Urban Planning for Small Cities in Egypt*, UN-Habitat, Earthscan, London, UK.
4. Khalil, H. (2012). Enhancing quality of life through strategic urban planning, *Sustainable Cities and Society*, Vol. 5, pp. 77–86.
5. Arandel, C. and El Batran, M. (1997). *The Informal Housing Development Process in Egypt*, Centre Nacional de la Recherche Scientifique, Bordeaux.
6. Khalil, H. (2012). Affordable housing: Quantifying the concept in the Egyptian context, *Journal of Engineering and Applied Science*, Vol. 59, No. 2, pp. 129–148.

6

Energy in Buildings

This chapter generally is devoted to defining energy-efficient buildings and the relevant indoor environmental quality; it gives a brief account of rating systems of energy-efficient buildings.

6.1 Energy-Efficient Buildings: A Challenging Era

The developing community in its path for rapid development is endeavouring to make all necessary and appropriate measures to enhance the efficiency of energy utilisation and to increase the beneficiation of the energy resources. Throughout Egypt, energy resources are widely used and consumption rates are in general exceeding the internationally accepted values. The use and application of new and renewable energy sources can be harnessed to design, construct, and to operate a solar building of moderate size for desert applications. This chapter demonstrates the importance of incorporating an energy performance directive as a standard in our region; such a goal will aid energy savings in large buildings and will set regulations for energy-efficient designs that are based on standard calculation methods. The target is to develop standardised tools for the calculation of the energy performance of buildings, with defined system boundaries for the different building categories and for different cooling/heating systems. We endeavour to prepare models for expressing requirements of indoor air quality, thermal comfort in winter (and, when appropriate, in summer), visual comfort and so on with a common procedure for an 'energy performance certificate'.

6.1.1 Energy Declaration of Buildings

All countries around the globe have developed or are now undertaking the development of the following specific codes with their compliance procedures and measures:

1. Residential Energy: Building Codes
2. Commercial Energy: Building Codes

The implementation of such a project is hoped to be realised through close cooperation and association with the Housing and Building National

Centre, universities, and the building and construction industry. The main targets of the Energy Performance Directive are

> 'Legislative authorities shall ensure that, when buildings are constructed, sold, or rented out, an energy performance certificate is made available to the owner or by the owner to the prospective buyer or tenant, as the case might be....'
>
> 'The energy performance certificate for buildings shall include reference values such as current legal standards and benchmarks in order to make it possible for consumers to compare and assess the energy performance of the building. The certificate shall be accompanied by recommendations for cost-effective improvement of the energy performance....' [Available at www.epbd.eu]

The following steps are required for the energy certification:

1. Develop methodologies for energy declaration of the buildings
2. Develop reference values (key numbers) and/or systems for benchmarking
3. Provide a labelling system for selected buildings

The primary use of energy declarations is, among others, to

1. Create consciousness of energy efficiency in buildings and also improve the knowledge of energy use in buildings
2. Use the information to determine if the building works as well as possible with regard to its technical design
3. Use the information for benchmarking
4. Use the information for suggesting measures and recommendations for reducing energy use
5. Provide the information necessary to make calculations of the environmental impact due to the energy use (e.g. carbon dioxide emissions)
6. Describe selected energy properties of the building
7. Give the basis for a common energy performance certification of a building

It is believed that depending on the purpose of the energy declaration, different procedures can be of interest. Different actors need different information. In order to give relevant advice to the property owner about which measures are cost-effective, a very careful examination and calculation of the building's energy balance is necessary. A careful analysis is also necessary to give relevant information to the users about how they can decrease their energy use without decreasing, to an unacceptable level, the indoor air quality and thermal comfort. The main purpose is to reduce the energy consumption in the commercial and public building sector and hence also to reduce carbon dioxide emissions. One way to proceed is to carry out the energy calculations in different steps for existing buildings.

1. The first step is to collect measured energy use information (e.g. from energy bills) and make a benchmarking to decide if the actual building is better or worse compared to similar buildings—for example, if the energy use seems to be higher than the average for a comparable grouping of buildings.

2. A second step is to make a careful energy calculation that can be compared to the measured energy use. This has to be carried out to identify the types of measures that can be recommended.

3. Some important aspects that are necessary to take into consideration when developing a common tool for energy declaration of buildings will be addressed. Figure 6.1 demonstrates the energy flow chart proposed in the European Energy Performance Directive [1,2].

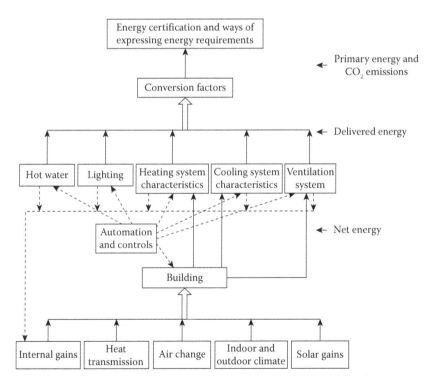

FIGURE 6.1
Energy flow chart in buildings. (From Khalil, E. E. (2005). Energy performance of buildings directive in Egypt: A new direction, *HBRC Journal*, Vol. 1.)

6.1.2 Energy Performance

The energy performance of buildings should include a general framework for the calculation of energy performance and the building categories to be included. Let us consider here the basic recommended method for calculating the energy performance of buildings [3–10]; it should include at least, but not limited to, the following aspects:

1. Thermal characteristics of the building (envelope and interior partitions and so on)
2. Air-conditioning system
3. Ventilation system
4. Hot water supply and winter seasonal heating
5. Built-in lighting installation for commercial sector
6. The building's orientation: its position and outdoor climatic conditions
7. Passive solar systems and solar protection
8. Natural ventilation (if available)
9. Indoor climatic design conditions

The positive influence of renewable energy sources on the environment should be included, such as

1. Active solar systems' contribution to domestic water heating based on renewable energy sources
2. Combined power and heat production
3. District cooling/heating systems
4. Natural lighting with goals to save electricity and cooling energy

Building categories to be covered can include

1. Office buildings
2. Educational facilities
3. Healthcare facilities and hospitals
4. Hotels and restaurants
5. Sporting facilities
6. Department stores and commercial buildings
7. Warehouses, museums, cold stores and so forth

Starting from the contents of the calculation framework and looking at the currently available European and international standards, Table 6.1 is the simple outcome summary. Preferably, as few calculation models as

TABLE 6.1

List of Energy-Related Standards and Aspects in Buildings

Aspect	Sub-Aspect	ISO Standards
Thermal characteristics	Building components	ISO 6946, 10292, 13370, 10077, 13789
Calculations of design heat load		ISO 13790, 13786
Air-conditioning installation	Cooling load, efficiency	ISO18618
Orientation of buildings, outdoor climate		ISO 13790, 15927
Climatic data		ISO 9050

possible should be used to cover the different building categories. It may, however, be necessary to use more than one model, depending on the desired output and accuracy.

A simplified calculation method for cooling load and cooling energy use including efficiency losses should be established. Work in this field has just started internationally. The terms 'air leakage', which is used in conjunction with the building envelope, and 'air infiltration', which is used in ventilation design, should be coordinated because they effectively mean the same thing. Areas where information on logistics is missing are those for natural and hybrid ventilation systems—where calculation guidelines are missing.

A comprehensive calculation method should clearly include

1. Method to build up internal gains by adding individual components from the bottom up.
2. Built-in lighting could be included, with the possibility of giving credit for natural lighting in this approach.
3. Calculation of energy supply from solar water heating and solar heating systems, including active seasonal storage or the use of other renewable energy sources such as wind and geothermal.
4. Indication of the bases for specific real energy use compared to design energy use.
5. Energy use normalisation criteria, such as kWh/m^2, for benchmarking purposes; a clear definition of which m^2 to be used is necessary.

6.1.3 Need for Further Development

The present work opens the field for further investigation to focus on some issues that have to be developed before the preparation and implementation of an energy directive for Egypt:

1. Cooling load and energy use for cooling.
2. If we assume lighting has high priority, a method to calculate all internal gains should be developed.

3. A method for establishing the air change rate based on air leakage (air infiltration), airing and natural/hybrid/mechanical ventilation should also be developed.

4. Verification and normalisation methods for credibility and to compare among countries.

5. Calculation schemes for the use of renewable energy sources in different applications in buildings and particularly in desert areas.

It is probable that one calculation method will not cover all aspects and building categories of the directive. For some applications, more advanced simulation models will have to be used to provide satisfactory accuracy. The on-going and future work on methods for validation and documentation of simulation tools [10,11] could be valuable in a process of approving models.

6.1.4 Mathematical Simulation Tools

A handful of calculation tools and simulation models have been developed over the past few years, both inside and outside of International Energy Agency (IEA) projects. These tools are still dispersed, and their integration into the global energy analysis in the design practice has not been achieved until now. In addition, a great deal of effort has also been invested by other research organisations to produce user-friendly simulation models; however, the practical use of existing simulation models is very limited in building and heating, ventilation, air conditioning (HVAC) design. Many powerful models stay unused because of their poor documentation, poor manuals, and/or because their users are not provided with simple and transparent procedures for model input generation. This can be improved by providing guidelines for input generation and for the use of standard product databases.

Simulation should be applied in all stages of the building design process so that quick feedback on design implications can be given. Simulation tools should also allow the designer to compare various options and see whether (or to what extent) each component is necessary. A well-defined task is to investigate the availability and accuracy of building energy analysis tools and engineering models to evaluate the performance of solar and low energy buildings. The scope of the task is limited to whole building energy analysis tools, including emerging modular type tools, and to widely used solar and low-energy design concepts. Tool evaluation activities will include analytical, comparative and empirical methods, with emphasis given to blind empirical validation using measured data from test rooms or full-scale buildings. Documentation of engineering models will use existing standard reporting formats and procedures. However, tool users, such as architects, air-conditioning engineers, energy consultants, product manufacturers, and building owners and managers, are the ultimate beneficiaries of such research and should be informed through targeted reports and articles.

A tool validation should proceed as follows:

1. Qualification of simulation models: The aim of this subtask is to make the simulations safer thanks to better selection and a better use of the models available. This activity will include a review of HVAC models available, a review of datasets, a definition of simulation requirements and a definition of qualification tests.

2. Design process analysis: Description of the building and energy data required for the use of simulation in the different phases of design, a description of simulation results to support the decisions in the different phases and a description of the information flow from preliminary design until audit and retrofit.

3. Information management: Assess the capabilities of the existing simulation tools [Department of Energy, EnergyPlus, American Society of heating, Refrigeration and Air Conditioning Engineers (ASHRAE)], establish a checklist of data required, propose default values and define possible support for data collection and communication language among users.

4. Data exchange: This includes a review of data exchange formats available and a selection of one data exchange format, which has a connection to a life cycle energy simulation programme.

6.1.5 Arab Energy in Buildings Code

A challenging task is to design, build, and set the guidelines for a reliable energy code that satisfies the general needs of the Arab nations and to tailor energy utilisation in buildings as follows:

1. Set minimum energy efficiency levels of general energy-consuming equipment in typical residential and/or commercial buildings.

2. Set the climatic data for the Arab world and build up the relevant building envelope data that are used for energy calculations based on local building materials and practices.

3. Set the prescriptive energy in kW/m^2 of floor area for different applications.

4. Devise a methodology of trade-off of energy-consuming elements to achieve an energy-efficient building.

5. Use internationally verified calculation tools of energy patterns and buildings' energy signatures such as those of the U.S. Department of Energy (DOE) [12].

6.1.6 Conclusions

From the aforementioned analyses, one may conclude the importance of incorporating an energy performance directive as a standard in our region; such a goal will aid energy savings in large buildings and will set regulations for energy-efficient designs that are based on standard calculation methods. The proposed standard would be largely based on international standards and appropriately modified to suit local practices. Basically, the proposal is to

1. Develop standardised tools for the calculation of the energy performance of buildings.
2. Define system boundaries for the different building categories.
3. Prepare models for expressing requirements on indoor air quality, thermal comfort in winter (and, when appropriate, in summer), visual comfort and so forth.
4. Develop transparent systems to determine necessary input data for the calculations, including default values on internal gains.
5. Define comparable energy-related key values (kWh/m^2, kWh per person, kWh per apartment, kWh per produced unit etc.). The areas/volumes need to be defined.
6. Develop a method to translate net energy used in the building to primary energy and carbon dioxide emissions.
7. Develop a common procedure for an 'energy performance certificate'.
8. Develop and compile relevant standards applicable for each individual building category.

6.2 Indoor Environmental Quality

6.2.1 General

Indoor environmental quality, as the name implies, simply refers to the quality of the air in an office or in other building environments. Workers are often concerned that they have symptoms or health conditions from exposure to contaminants in the buildings where they work. One reason for this concern is that their symptoms often get better when they are not in the building. Although research has shown that some respiratory symptoms and illnesses can be associated with damp buildings, it is still unclear what measurements of indoor contaminants show that workers are at risk for disease. In most instances, where a worker and his or her physician suspect that the building environment is causing a specific health condition, the information available from medical tests and tests of the environment is not sufficient to establish which contaminants are responsible. Indoor environments are highly complex, and building occupants may be exposed to a variety of contaminants (in the form of gases and particles) from office machines, cleaning products, construction activities, carpets and furnishings, perfumes, cigarette smoke, water-damaged building materials, microbial growth (fungal/mould and bacterial), insects and outdoor pollutants. Other factors such as indoor temperatures, relative humidity and ventilation levels can also affect how individuals respond to the indoor environment. Understanding the sources of indoor environmental contaminants and controlling them can often help prevent or resolve building-related worker symptoms. Practical guidance for improving and maintaining the indoor environment is available. Workers who have persistent or worsening symptoms should seek medical evaluation to establish a diagnosis and to obtain recommendations for treatment of their condition.

This section specifies requirements for indoor environmental quality including indoor air quality, environmental tobacco smoke control, outdoor air delivery monitoring, thermal comfort, building entrances, acoustic control, day lighting and low-emitting materials.

6.2.2 Definitions

The following words and terms shall, for the purposes of this chapter and as used elsewhere in this code, have the meanings shown herein:

1. Agrifibre products: Agrifibre products include wheat board, strawboard, panel substrates and door cores, but do not include furniture, fixtures and equipment (FF&E) not considered base building elements.

2. Composite wood products: Composite wood products include hardwood plywood, particleboard and medium-density fibreboard. Composite wood products do not include hardboard, structural plywood, structural panels, structural composite lumber, oriented strand board, glued laminated timber as specified in 'Structural Glued Laminated Timber' (ANSI A190.1-2002) or prefabricated wood I-joists.

3. Daylit area: Area under horizontal fenestration (skylight) or adjacent to vertical fenestration (window) and described in Figures 6.2 and 6.3.

4. Daylit space: Space bounded by vertical planes rising from the boundaries of the daylit area on the floor to above the floor or roof.

5. HVAC units, small: Those containing less than 0.22 kg of refrigerant.

6. Interior, building: The inside of the weatherproofing system.

7. MERV: Filter minimum efficiency reporting value, based on ASHRAE 52.2-2009.

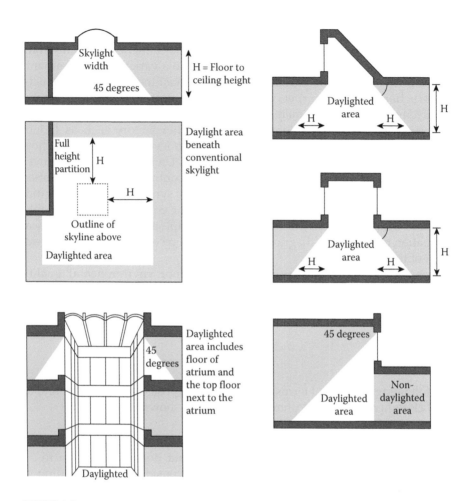

FIGURE 6.2
Horizontal daylight area.

8. Moisture content: The weight of the water in wood expressed as a percentage of the weight of the oven-dry wood.

9. Multi-occupant spaces: Indoor spaces used for presentations and training, including classrooms and conference rooms.

10. Single occupant spaces: Private offices, workstations in open offices, reception work stations and ticket booths.

11. VOC: A volatile organic compound broadly defined as a chemical compound based on carbon chains or rings with vapour pressures greater than 0.1 mm of mercury at room temperature. These compounds typically contain hydrogen and may contain oxygen, nitrogen and other elements.

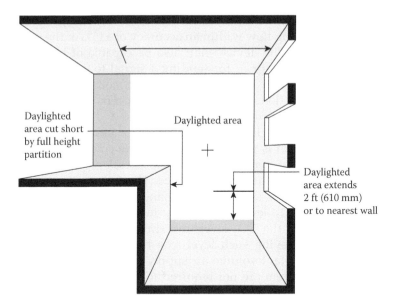

FIGURE 6.3
Vertical daylight area.

6.2.3 Requirements

This section specifies requirements for indoor environmental quality including indoor air quality, environmental tobacco smoke control, outdoor air delivery monitoring, thermal comfort, building entrances, acoustic control, day lighting and low-emitting materials. A typical building should comply with the indoor air quality requirements as outlined in international codes and standards (ASHRAE Standard 62.1) and ISO 16818.

6.2.4 Minimum Ventilation Rates

1. The ventilation rate procedure outlined in ASHRAE Standard 62.1 is recommended as a tool to design each mechanical ventilation system in the building.
2. The zone-level design outdoor airflow rates in all occupiable spaces shall be greater than or equal to the airflow calculated using the ventilation rate procedure.

6.2.5 Outdoor Air Delivery Monitoring

6.2.5.1 Spaces Ventilated by Mechanical Systems

For variable air volume supply systems, a permanently mounted, direct total outdoor airflow measurement device shall be provided that is capable of measuring the system's minimum outdoor airflow rate. The device shall be capable of measuring flow within an accuracy of ±15% of the minimum outdoor airflow rate. The device shall also be capable of being used to warn the building operator or for sending a signal to a building central monitoring system when flow rates are not in compliance. Naturally, there are some exceptions to this requirement. For each air handling system that serves only *densely occupied spaces*, a direct total outdoor airflow measurement device is not required if a permanently installed CO_2 monitoring system is provided for all *densely occupied spaces*. For air handling systems that serve only a single *densely occupied space*, CO_2 monitoring is required only for that space. The CO_2 monitoring system shall record ventilation system performance in terms of differential indoor-to-outdoor CO_2 levels. The CO_2 monitoring system shall be capable of indicating the CO_2 level in a direct read-out display in the occupied space, conveying such level to a building central monitoring system, or both. Constant volume air supply systems that use a damper position feedback system are not required to have a direct total outdoor airflow measurement device.

6.2.5.2 Naturally Ventilated Spaces

A permanently installed CO_2 monitoring system shall be provided in *densely occupied spaces* designed to operate without a mechanical ventilation system for any period of time that the space is occupied. Indoor CO_2 sensors or air sampling probes shall be located within the room between 1 and 2 m above the floor and on a wall location at least 6 m from operable openings. The CO_2 monitoring system shall be capable of indicating the CO_2 level in a direct read-out display in the occupied space, conveying such level to a building central monitoring system, or both. Where floor plans are less than 12 m wide, sensors shall be located as close to the centre of the space as practical. One CO_2 sensor is allowed to be used to represent multiple spaces if the natural ventilation design uses passive stack(s) or other means to induce air-flow through those spaces equally and simultaneously without intervention by building occupants.

6.2.5.3 CO₂ Sensors

Spaces with CO_2 sensors or air sampling probes leading to a central CO_2 monitoring station shall have one sensor or probe for each 1000 m² of floor space and shall be located in the room between 1 and 2 m above the floor. CO_2 sensors must be accurate to ±50 ppm at 1000 ppm. For all spaces with CO_2 sensors, *target concentrations* shall be calculated for full and part load conditions (to include no less than 25%, 50%, 75%, and 100%). Metabolic rate, occupancy, ceiling height, and *outdoor air* CO_2 concentration assumptions and *target concentrations* shall be shown in the design documents. *Outdoor air* CO_2 concentrations shall be determined by one of the following:

1. CO_2 concentration shall be assumed to be 400 ppm without any direct measurement
2. CO_2 concentration shall be dynamically measured using a CO_2 sensor located near the position of the *outdoor air* intake

6.2.6 Filtration and Air Cleaner Requirements

6.2.6.1 Particulate Matter

1. The particulate matter filters or air cleaners shall have an Minimum Energy Values (MERV) of not less than 8 and shall comply with and be provided where required in Section 5.9 of ASHRAE Standard 62.1.
2. In addition to ASHRAE Standard 62.1 Section 6.2.1.1, when the building is located in an area that is designated 'non-attainment' with the National Ambient Air Quality Standards for PM2.5 as determined by the American Heating Institute (AHI) (in the United States by the Environmental Protection Agency [EPA]), particle filters or air-cleaning devices having a MERV of not less than 13 when rated in accordance with ASHRAE Standard 52.2 shall be provided to clean outdoor air prior to its introduction to occupied spaces.

6.2.6.2 Ozone

In addition to ASHRAE Standard 62.1 Section 6.2.1.2 requirements, when the building is located in an area that is designated 'non-attainment' with the National Primary and Secondary Ambient Air Quality Standards for ozone as determined by the AHJ (in the United States, by the EPA), air cleaner cleaning devices having a removal efficiency of no less than the one efficiency specified in ASHRAE Standard 62.1 Section 6.2.1.2 shall be provided to clean outdoor air prior to its introduction to occupied spaces.

6.2.6.3 Bypass Pathways

All filter frames, air cleaner racks, access doors and air cleaner cartridges shall be sealed to eliminate air bypass pathways.

6.2.7 Building Requirements

6.2.7.1 Building Entrances

All *building entrances* shall employ an entry mat system that shall have a scraper surface, an absorption surface and a finishing surface. Each surface shall be a minimum of the width of the entry opening, and the minimum length is measured in the primary direction of travel. Exceptions can be made for

1. Entrances to individual *dwelling units.*
2. Length of entry mat surfaces is allowed to be reduced due to a barrier, such as a counter, partition or wall, or due to local regulations prohibiting the use of scraper surfaces outside the entry. In this case, entry mat surfaces shall have a minimum length of 1 m of indoor surface, with a minimum combined length of 2 m.

6.2.7.1.1 Scraper Surface

The scraper surface shall comply with the following:

1. Shall be the first surface stepped on when entering the building
2. Shall be either immediately outside or inside the entry
3. Shall be a minimum of 1 m long
4. Shall be either permanently mounted grates or removable mats with knobby or squeegee-like projections

6.2.7.1.2 Absorption Surface

The absorption surface shall comply with the following:

1. Shall be the second surface stepped on when entering the building
2. Shall be a minimum of 1 m long and made from materials that can perform both a scraping action and a moisture wicking action

6.2.7.1.3 Finishing Surface

The finishing surface shall comply with the following:

1. Shall be the third surface stepped on when entering the building
2. Shall be a minimum of 1.2 m long and made from material that will both capture and hold any remaining particles or moisture

6.2.8 Thermal Environmental Conditions for Human Occupancy Comfort

The building shall be designed in compliance with ASHRAE Standard 55. Exceptions are spaces with special requirements for processes, activities or contents that require a thermal environment outside that which humans find thermally acceptable, such as food storage, natatoriums, shower rooms, saunas and drying rooms.

6.2.9 Acoustical Control

6.2.9.1 Exterior Sound

Wall and roof-ceiling assemblies that are part of the *building envelope* shall have a composite outdoor–indoor transmission class (OITC) rating of 40 or greater or a composite sound transmission class (STC) rating of 50 or greater, and *fenestration* that is part of the *building envelope* shall have an OITC or STC rating of 30 or greater for any of the following conditions:

1. Buildings within 300 m of *expressways*
2. Buildings within 8 km of airports serving more than 10,000 commercial jets per year
3. Where *yearly average day-night average sound levels* at the property line exceed 65 decibels

Exceptions to that are buildings that may have to adhere to functional and operational requirements such as factories, stadiums, storage, enclosed parking structures and utility buildings.

6.2.9.2 Interior Sound

Interior *wall* and floor/ceiling assemblies separating interior rooms and spaces shall be designed in accordance with all of the following:

1. *Wall* and floor-ceiling assemblies separating adjacent *dwelling units*, *dwelling units* and public spaces, adjacent tenant spaces, tenant spaces and public places, and adjacent *classrooms* shall have a composite STC rating of 50 or greater.
2. *Wall* and floor-ceiling assemblies separating hotel rooms, motel rooms and patient rooms in nursing homes and hospitals shall have a composite STC rating of 45 or greater.
3. *Wall* and floor-ceiling assemblies separating *classrooms* from rest rooms and showers shall have a composite STC rating of 53 or greater.
4. *Wall* and floor-ceiling assemblies separating *classrooms* from music rooms, mechanical rooms, cafeteria, gymnasiums and indoor swimming pools shall have a composite STC rating of 60 or greater.

6.2.9.3 OITC and STC

OITC values for assemblies and components shall be determined in accordance with ASTM E1332. STC values for assemblies and components shall be determined in accordance with ASTM E90 and ASTM E413.

6.2.10 Day Lighting by Top Lighting

There shall be a minimum *fenestration area* providing day lighting by *top lighting* for large enclosed spaces. In buildings three stories and less above grade, conditioned or unconditioned enclosed spaces that are greater than 2000 m² directly under a *roof* with a finished ceiling height greater than 4 m and that have a *lighting power allowance* for general lighting equal to or greater than 5.5 W/m², shall comply with the following.

6.2.10.1 Minimum Daylight Zone by Top Lighting

A minimum of 50% of the floor area directly under a *roof* in spaces with a lighting power density or *lighting power allowance* greater than 5 W/m² shall be in the *daylight zone*. Areas that are daylit shall have a minimum *top lighting* area to *daylight zone* area ratio as shown in Table 6.2. For purposes of compliance with Table 6.2, the greater of the space lighting power density and the space *lighting power allowance* shall be used.

6.2.10.2 Skylight Characteristics

Skylights used to comply with Section 6.2.10.1 shall have a glazing material or diffuser that has a measured haze value greater than 90%, tested according to ASTM D1003 (notwithstanding its scope) or another test method approved by the *authority having jurisdiction*. Exceptions are for *skylights* with a measured haze value less than or equal to 90% whose combined area does not exceed 5% of the total *skylight* area.

   Skylights are capable of preventing direct sunlight from entering the occupied space below the well during occupied hours. This shall be accomplished using one or more of the following:

1. Orientation
2. Automated shading or diffusing devices
3. Diffusers
4. Fixed internal or external baffles

TABLE 6.2

Minimum Top Lighting Area

General Lighting Power Density or Lighting Power Allowances in Daylight Zone	
W/m²	Minimum Top Lighting Area to Daylight Zone Area Ratio
14 W/m² < LPD	3.6%
10 W/m² < LPD < 14 W/m²	3.3%
5 W/m² < LPD < 10 W/m²	3.0%

6.2.11 Isolation of the Building from Pollutants in Soil

For building projects that include construction or expansion of a ground-level foundation and that are located on 'brownfield' sites or in 'Zone 1' and are located in counties identified to have a significant probability of radon concentrations higher than 4 picocuries/L on the U.S. EPA Map of Radon Zones, shall have a *soil gas retarder system* installed between the newly constructed space and the soil.

6.2.12 Prescriptive Option

6.2.12.1 Day Lighting by Side Lighting

6.2.12.1.1 Minimum Effective Aperture

Office spaces and *classrooms* shall comply with the following criteria: All north-, south- and east-facing facades for those spaces shall have a minimum *effective aperture for vertical fenestration*. Opaque interior surfaces in *daylight zones* shall have visible light reflectance greater than or equal to 80% for ceilings and 70% for partitions higher than 1.5 m in *daylight zones*.

6.2.12.1.2 Office Space Shading

Each west-, south- and east-facing facade shall be designed with a shading *projection factor*. Shading is allowed to be external or internal using the *projection factor, interior*. The building is allowed to be rotated up to 45 degrees for purposes of calculations and for showing compliance. The following shading devices are allowed to be used: louvers, sun shades, light shelves, and any other permanent device. Any *vertical fenestration* that employs a combination of interior and external shading is allowed to be separated into multiple segments for compliance purposes. Each segment shall comply with the requirements for either external or *interior projection factor*.

1. Building self-shading through *roof* overhangs or recessed windows.
2. External buildings and other permanent infrastructure or geological formations that are not part of the building. Trees, shrubs or any other organic shading device shall not be used to comply with the shading *projection factor* requirements. Exceptions to that include *building projects* that comply with the prescriptive compliance option: *Vertical fenestration* that receives direct solar radiation for less than 250 hours per year because of shading by permanent external buildings, existing permanent infrastructure, or topography.

6.2.13 Materials

Reported emissions or VOC contents specified here shall be from a representative product sample and conducted with each product reformulation or at a minimum of every three years. Products certified under third-party certification programmes as meeting the specific emission or VOC content requirements listed here are exempted from this three-year testing requirement but shall meet all the other requirements as listed here.

6.2.13.1 Adhesives and Sealants

Products in this category include carpet, resilient, and wood flooring adhesives; base cove adhesives; ceramic tile adhesives; drywall and panel adhesives; aerosol adhesives; adhesive primers; acoustical sealants; fire stop sealants; HVAC air duct sealants; sealant primers; and caulks. All adhesives and sealants used on the interior of the building (defined as inside of the *weatherproofing system* and applied on-site) shall comply with the requirements of either 6.2.13.1.1 or 6.2.13.1.2.

6.2.13.1.1 Emissions Requirements

Emissions shall be determined according to CA/DHS/EHLB/R-174 (commonly referred to as California Section 01350) and shall comply with the limit requirements for either office or *classroom* spaces regardless of the space type.

6.2.13.1.2 VOC Content Requirements

VOC content shall be complying with and shall be determined according to and complying with the following limit requirements:

1. Adhesives, sealants, and sealant primers: SCAQMD Rule 1168. HVAC duct sealants shall be classified as in the 'Other' category within the SCAQMD Rule 1168 sealants table.
2. Aerosol adhesives: Green Seal Standard GS-36.

Exceptions to 6.2.13.1: The following solvent welding and sealant products are not required to meet the emissions or the aforementioned VOC content requirements: Cleaners, solvent cements, and primers used with plastic 'piping and conduit in plumbing, fire suppression, and electrical systems. HVAC air duct sealants when the air temperature of the space in which they are applied is less than 4.5°C.

6.2.13.1.3 Paints and Coatings

Products in this category include sealers, stains, clear wood finishes, floor sealers and coatings, waterproofing sealers, primers, flat paints and coatings, non-flat paints and coatings, and rust preventative coatings.

Paints and coatings used on the interior of the building (defined as inside of the *weatherproofing system* and applied on-site) shall comply with either 6.2.13.2.1 or 6.2.13.2.2.

6.2.13.2 Emissions

6.2.13.2.1 Emissions Requirements

Emissions shall be determined according to CA/DHS/EHLB/R-174 (commonly referred to as California Section 01350) and shall comply with the limit requirements for either office or classroom spaces regardless of the space type.

6.2.13.2.2 VOC Content Requirements

VOC content shall comply with and be determined according to and comply with the following limit requirements:

1. Architectural paints, coatings and primers applied to interior surfaces *walls* and ceilings: Green Seal Standard GS-11
2. Clear wood finishes, floor coatings, stains, sealers and shellacs: SCAQMD rules

6.2.13.3 Floor Covering Materials

Floor covering materials installed in the building interior shall comply with the following:

1. Carpet: Carpet shall be tested in accordance with and shown to be compliant with the requirements of CA/DHS/EHLB/R-174 (commonly referred to as California Section 01350). Products that have been verified and labelled to be in compliance with Section 9 of the CA/DHS/EHLB/R-174 comply with this requirement.
2. Hard surface flooring in office spaces and *classrooms*: Materials shall be tested in accordance with and shown to be compliant with the requirements of CA/DHS/EHLB/R-174 (commonly referred to as California Section 01350). Products that have been verified and labelled to be in compliance with SCS-EC10.2 comply with this requirement.

6.2.13.4 Composite Wood, Wood Structural Panel and Agrifibre Products

Composite wood, wood structural panels and agrifibre products used on the interior of the building (defined as inside of the *weatherproofing system*) shall contain no added urea–formaldehyde resins. Laminating adhesives used to fabricate on-site and shop-applied composite wood and agrifibre assemblies shall contain no added urea–formaldehyde resins. Composite wood and agrifibre products are defined as: particleboard, medium-density fibreboard, wheat board, strawboard, panel substrates and door cores. Materials considered as FF&E are not considered base building elements and are not included in this requirement. Emissions for products covered by this section shall be determined according to and shall comply with one of the following:

1. Third-party certification shall be submitted indicating compliance with the California Air Resource Board's (CARB) regulation entitled: *Airborne Toxic Control Measure to Reduce Formaldehyde Emissions from Composite Wood Products*; third-party certifiers shall be approved by CARB.
2. CA/DHS/EHLB/R-174 (commonly referred to as California Section 01350) and shall comply with the limit requirements for either office or classroom spaces regardless of the space type.

Exceptions are for structural panel components such as plywood, particle board, wafer board and oriented strand board identified as 'EXPOSURE 1', 'EXTERIOR' or 'HUD-APPROVED' are considered acceptable for interior use.

6.2.13.5 Office Furniture Systems and Seating

All office furniture systems and seating installed prior to occupancy shall be tested according to ANSI/BIFMA Standard M7.1 testing protocol and shall not exceed the limit requirements of this standard.

6.2.13.6 Ceiling and Wall Systems

These systems include ceiling and wall insulation, acoustical ceiling panels, tackable wall panels, gypsum wall board and panels, and wall coverings. Emissions for these products shall be determined according to CA/DHS/EHLB/R-174 (commonly referred to as California Section 01350) and shall comply with the limit requirements for either office or classroom spaces regardless of the space type.

6.2.14 Performance Option

6.2.14.1 Day Lighting Simulation

6.2.14.1.1 Usable Illuminance in Office Spaces and Classrooms

The design for the *building project* shall demonstrate an illuminance of at least 30 fc (300 lux) on a plane 1 m above the floor, within 75% of the area of the *daylight zones.* The simulation shall be made at noon on the equinox using an accurate physical or computer day lighting model.

1. Computer models shall be built using daylight simulation software based on the ray tracing or radiosity methodology.
2. Simulation is to be done using either the CIE Overcast Sky Model or the CIE Clear Sky Model.

Exception to 6.2.14.1.1 is where the simulation demonstrates that existing adjacent structures preclude meeting the illuminance requirements.

6.2.14.2 Direct Sun Limitation on Work Plane Surface in Offices

It shall be demonstrated that direct sun does not strike anywhere on a work surface in the work plane in any daylit space for more than 20% of the occupied hours during an equinox day in regularly occupied office spaces. If the work surface height is not defined, a height of 0.75 m above the floor shall be used.

6.2.14.3 Materials

The emissions of all the materials listed here and used within the building (defined as inside of the *weatherproofing system* and applied on-site) shall be modelled for individual VOC concentrations. The sum of each individual VOC concentration from the materials listed here shall be shown to be in compliance with the limits as listed in Section 4.3 of the CA/DHS/EHLB/R-174 (commonly referred to as California Section 01350) and shall be compared to 100% of its corresponding listed limit. In addition, the modelling for the building shall include at a minimum criteria. Emissions of materials used for modelling VOC concentrations shall be obtained in accordance with the testing procedures of CA/DHS/EHLB/R-174 unless otherwise noted below.

1. Tile, strip, panel and plank products including vinyl composition tile, resilient floor tile, linoleum tile, wood floor strips, parquet flooring, laminated flooring and modular carpet tile.
2. Sheet and roll goods including broadloom carpet, sheet vinyl, sheet linoleum, carpet cushion, wall covering and other fabrics.
3. Rigid panel products including gypsum board, other wall panelling, insulation board, oriented strand board, medium density fibre board, wood structural panel, acoustical ceiling tiles and particleboard.
4. Insulation products.
5. Containerised products including adhesives, sealants, paints, other coatings, primers and other 'wet' products.

6. Cabinets, shelves and work surfaces that are permanently attached to the building before occupancy. Emissions of these items shall be obtained in accordance with the ANSI/BIFMA Standard M7.1.

7. Office furniture systems and seating installed prior to initial occupancy. Emissions from these items shall be obtained in accordance with the ANSI/BIFMA Standard M7.1.

Exception to 6.2.14.2 are salvaged materials that have not been refurbished or refinished within one year prior to installation.

6.3 Rating Systems of Energy-Efficient Buildings

6.3.1 Introduction

Attempts to adequately design an optimum HVAC airside system that furnishes comfort and air quality in air-conditioned spaces with efficient energy consumption comprise is a great challenge. Air conditioning identifies the conditioning of air for maintaining specific conditions of temperature, humidity and dust levels inside an enclosed space. The conditions to be maintained are dictated by the need for which the conditioned space is intended and for the comfort of users. So, the air conditioning embraces more than cooling or heating. The comfort air conditioning is defined as 'the process of treating air to control simultaneously its temperature, humidity, cleanliness and distribution to meet the comfort requirements of the occupants of the conditioned space' [13]. Air conditioning, therefore, includes the entire heat exchange operation as well as the regulation of velocity, thermal radiation and quality of air, as well as the removal of foreign particles and vapours. Achieving occupant comfort and health is the result of a collaborative effort of environmental conditions, such as indoor air temperature, relative humidity, airflow velocity, pressure relationship, air movement efficiency, contaminant concentration, illumination and visual comfort, sound and noise, and other factors. In the holistic approach, the totality of the effects of the heat sink and sources in the building and the technical building systems that are recoverable for space conditioning are typically considered in the calculation of the thermal energy needs.

Because the technical building thermal system losses depend on the energy input, which itself depends on the recovered system thermal sources, iteration might be required.

The calculation procedure can be devised as follows:

1. Subsystem calculations are first performed as per prevailing standards and that will be followed by determination of the recoverable thermal system losses.

2. The recoverable thermal system losses are then added to the other recoverable heat sources already included (e.g. solar and internal heat gains, recoverable thermal losses from lighting and/or other technical building systems such as domestic hot water) in the calculation of the needs for heating and cooling.

3. Thermal energy needs for heating and cooling are then recalculated.

4. Iterations are performed from Step 1 until the calculated changes of the energy needs between two successive iterations are less than a defined limit (e.g. 1%) or calculations will be stopped after a specified number of iterations.

5. Continue to calculate the difference between the energy at the start of the iterations and the end; these are the recovered system thermal losses.

Examples of the internationally recognised energy labels are listed here. These are summarised in terms of applications in buildings, both residential and commercial. These include

1. Residential washing machines
2. Household refrigerators
3. Residential air conditioners
4. Water heaters

6.3.2 Major Appliances

For major appliances, energy labels are separated into at least four categories:

1. The appliance's details: according to each appliance, specific details of the model and its materials.

2. Energy class: a colour code associated with a letter (from A to G) that gives an idea of the appliance's electrical consumption.

3. Consumption, efficiency, capacity and so on: this section gives information according to appliance type.

4. Noise: the noise emitted by the appliance is described in decibels.

6.3.2.1 European Union Energy Label

According to several different European Union (EU) directives (92/75/CEE, 94/2/CE, 95/12/CE, 96/89/CE, 2003/66/CE, Italia), most white goods, light bulb packaging and cars must have an EU energy label (see Figure 6.4) clearly displayed when these items are offered for sale or rent. The energy efficiency of the appliance is rated in terms of a set of energy efficiency

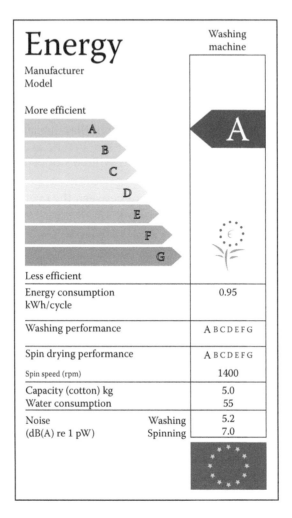

FIGURE 6.4
Example European Union energy label for a washing machine.

classes from A to G (on the label), with A being the most energy efficient and G the least efficient. The labels also give other useful information to the customer as they choose among various models. The information should also be given in catalogues and be included by Internet retailers on their websites.

In an attempt to keep up with advances in energy efficiency, A+ and A++ grades were later introduced for refrigeration products. Following public consultation from 2007–2008, a motion was put forward and eventually passed to introduce A+, A++ and A+++ grades for all appliances in order to bring the labelling system up to date by early 2011 [3].

6.3.2.1.1 Refrigerators, Freezers and Combined Appliances

The energy efficiency labels for freezers and refrigerators are shown in Figure 6.5; the index is calculated for each appliance according to its consumption and its compartments' volume, taking into account the appliance type. The index is thus not calculated in KWh.

The label also contains:

1. The annual energy consumption in KWh per year
2. The capacity of fresh foods in litres for refrigerators and combined appliances
3. The capacity of frozen foods in litres for freezers and combined appliances
4. The noise in dB(A)

For cold appliances (and this product alone), for models that are more economical than those of category A, categories A+ and A++ have been assigned.

6.3.2.1.2 Washing Machines, Tumble Dryers and Combined Appliances

For washing machines, the energy efficiency scale is calculated using a cotton cycle at 60°C with a maximum declared load (as shown in Figure 6.6). This load is typically 6 kg. The energy efficiency index is in KWh per kilograms of washing.

The energy label also contains information on

1. Total consumption per cycle
2. Washing performance, with a class from A to G
3. Spin drying performance, with a class from A to G
4. Maximum spin speed
5. The total cotton capacity in kg

A++	A+	A	B	C	D	E	F	G
<30	<42	<55	<75	<90	<100	<110	<125	>125

FIGURE 6.5
Label for refrigerators and freezers.

A	B	C	D	E	F	G
<0.19	<0.23	<0.27	<0.31	<0.35	<0.39	>0.39

FIGURE 6.6
Label for washing machines and tumble dryers.

6. Water consumption per cycle in litres

7. Noise in the washing and spinning cycles in dB(A)

For tumble dryers, the energy efficiency scale is calculated using the cotton drying cycle with a maximum declared load. The energy efficiency index is in KWh per kilograms of load. Different scales apply for condenser and vented dryers (as shown here in Figures 6.7 and 6.8).

The label also contains:

1. The energy consumption per cycle

2. The total cotton capacity

3. Whether the unit is vented or condensing

4. Noise in dB(A)

For combined washer dryers, the energy efficiency scale is calculated using the cotton drying cycle with a maximum declared load. The energy efficiency index is in KWh per kilograms of load. Different scales apply for condenser and vented dryers (shown in Figure 6.9).

The label also contains:

1. The energy consumption per cycle (washing and drying)

2. The energy consumption per cycle (washing only)

3. Washing performance, with a class from A to G

4. The maximum spin speed

5. The total cotton capacity (washing and drying separately)

6. Water consumption for a full load washed and dried; note that condenser dryers may use significant amounts of water on the drying cycle

7. Noise in dB (A) (separately for washing, spinning and drying)

A	B	C	D	E	F	G
<0.55	<0.64	<0.73	<0.82	<0.91	<1.00	>1.00

FIGURE 6.7
Label for condenser dryers.

A	B	C	D	E	F	G
<0.51	<0.59	<0.67	<0.75	<0.83	<0.91	>0.91

FIGURE 6.8
Label for vented dryers.

A	B	C	D	E	F	G
<0.68	<0.81	<0.93	<1.05	<1.17	<1.29	>1.29

FIGURE 6.9
Label for combined washer/dryers.

A	B	C	D	E	F	G
<1.06	<1.25	<1.45	<1.65	<1.85	<2.05	>2.05

FIGURE 6.10
Label for dishwashers.

6.3.2.1.3 Dishwashers

For dishwashers, the energy efficiency is calculated according to the number of place settings. For the most common size of appliance, the 12 place setting machine, the following classes apply as shown in Figure 6.10. The unit is expressed in KWh per 12 place settings.

The label also contains:

1. The energy consumption in kWh per cycle
2. The efficiency of the washing cycle, with a class from A to G
3. The efficiency of the drying cycle, with a class from A to G
4. The capacity as a number of place settings
5. The water consumption in litres per cycle
6. Noise in dB(A)

6.3.2.1.4 Ovens

The label also contains:

1. The efficiency, with a class from A to G
2. The energy consumption in kWh
3. The volume in litres
4. The type (small/medium/large)

6.3.2.1.5 Air Conditioners

The directive applies only to units under 12 kW. On every label, one shall find:

1. The model
2. The energy efficiency category from A to G
3. The annual energy consumption (full load at 500 hours per year)

	A	B	C	D	E	F	G
Cooling EER W/W	>3.2	3.0–3.2	2.8–3.0	2.6–2.8	2.4–2.6	2.2–2.4	<2.2
Heating COP W/W	>3.6	3.4–3.6	3.2–3.4	2.8–3.2	2.6–2.8	2.4–2.6	<2.4

FIGURE 6.11
Label for air conditioners.

4. The cooling output at full load in kW
5. The energy efficiency ratio (EER) in cooling mode at full load
6. The appliance type (cooling only, cooling/heating)
7. The cooling mode (air or water cooled)
8. The noise rating in dB (where applicable)

For air conditioners with heating capability, one shall also find (Figure 6.11):

1. The heat output at full load in kW
2. The heating mode energy efficiency category

Note that there exist units with EER and coefficient of performance (COP) > 5, so take a note of the actual number when it is A rated.

6.3.2.1.6 *Light Bulbs*

On every label, you will find:

1. The energy efficiency category from A to G
2. The luminous flux of the bulb in lumens
3. The electricity consumption of the lamp in watts
4. The average life length in hours

6.3.2.1.7 *Televisions*

New standards that televisions need to meet for the Energy Star efficiency rating came into effect early in September 2009 [4].

6.3.2.2 U.K. Energy Performance Certificate

A sample is shown here in Figure 6.12.

6.3.2.2.1 *Home Energy Performance Rating Charts*

Energy performance certificates (EPCs) are part of home information packs [1], which have been in effect since August 2007 in England and Wales for domestic properties with four or more bedrooms. The scheme was extended to encompass three-bedroom homes from September 2007. Rental properties have a certificate valid for 10 years that is required on a new tenancy commencing on

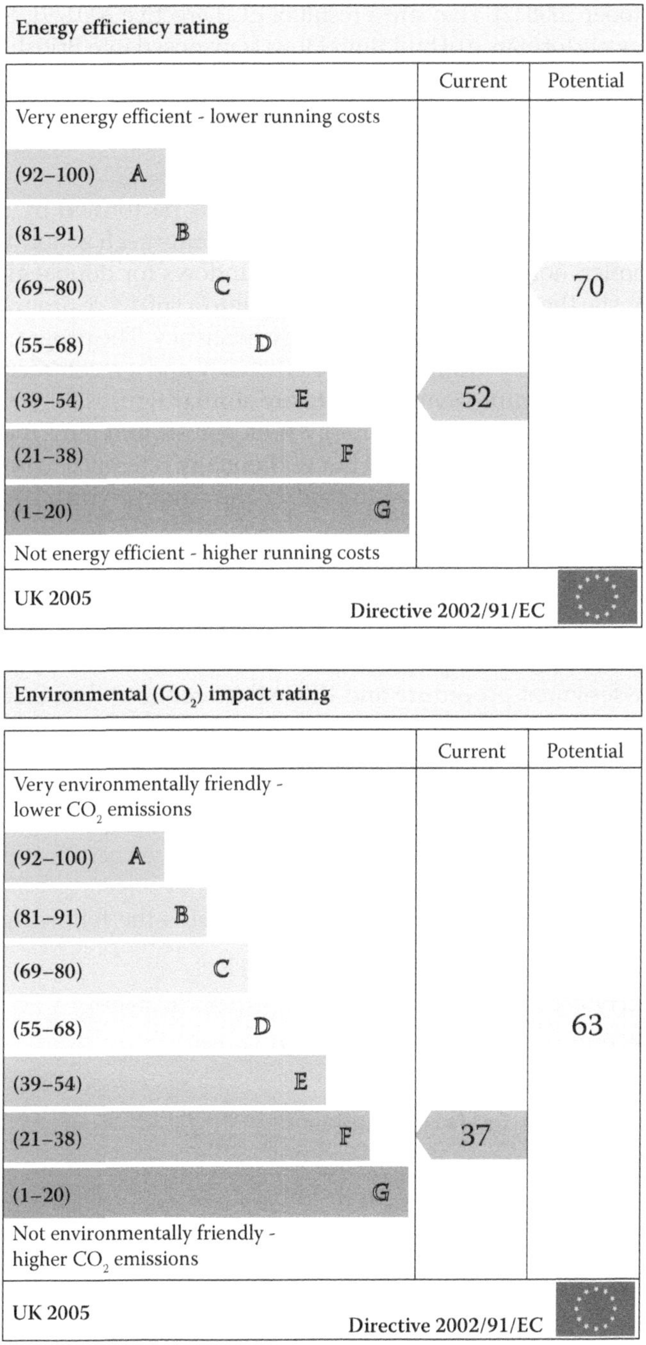

Energy efficiency rating

	Current	Potential
Very energy efficient - lower running costs		
(92–100) A		
(81–91) B		
(69–80) C		70
(55–68) D		
(39–54) E	52	
(21–38) F		
(1–20) G		
Not energy efficient - higher running costs		
UK 2005	Directive 2002/91/EC	

Environmental (CO$_2$) impact rating

	Current	Potential
Very environmentally friendly - lower CO$_2$ emissions		
(92–100) A		
(81–91) B		
(69–80) C		
(55–68) D		63
(39–54) E		
(21–38) F	37	
(1–20) G		
Not environmentally friendly - higher CO$_2$ emissions		
UK 2005	Directive 2002/91/EC	

FIGURE 6.12
Example of energy label for United Kingdom.

or after October 2008 [2]. They are a result of EU Directive 2002/91/EC relating to the energy performance of buildings [3], as transposed into British law by the Housing Act 2004 and The Energy Performance of Buildings (Certificates and Inspections) (England and Wales) Regulations 2007 [4].

6.3.2.2.2 Procedure

The energy survey needed to produce an EPC is performed by an inspector who visits the property and examines key items such as loft insulation, domestic boiler, hot water tank, radiators, windows for double glazing and so on. He or she then inputs the observations into a software programme that actually performs the calculation of energy efficiency. The programme gives a single number for the rating of energy efficiency and a recommended value of the potential for improvement. There are similar figures for environmental impact. A table of estimated energy bills per annum (and the potential for improvement) is also presented but without any reference to householder bills. The householder will have to pay for the survey, which costs around £60 for a four-bedroom house.

6.3.2.2.3 Domestic EPCs

The calculation of the energy ratings on the EPC is based on the RDSAPv3 procedure, which is a simplified version of the SAP2005. SAP is short for standard assessment procedure and RDSAP for reduced data SAP; both are derived from the UK Building Research Establishment's Domestic Energy Model (BREDEM), which was originally developed in the 1980s and also underlies the National Home Energy Rating (NHER) rating. EPCs have to be produced as part of a home information pack by home inspectors or domestic energy assessors who are registered under an approved certification scheme.

6.3.2.2.3.1 Property Details　　The certificate contains the following property details [5]:

1. Property address
2. Property type (for example, detached house)
3. Date of inspection
4. Certificate date and serial number
5. Total floor area

The *total floor area* is the area contained within the external walls of the property. The figure includes internal walls, stairwells, and the like, but it excludes garages, porches, areas less than 1.5 m high, balconies and any similar area that is not an internal part of the dwelling [6].

6.3.2.2.3.2 The A to G Scale　　EPCs present the energy efficiency of dwellings on a scale of A to G. The most efficient homes—those which should have the lowest fuel bills—are in band A. The certificate uses the same scale to define the impact a home has on the environment. Better-rated homes should have less impact through carbon dioxide emissions. The average property in the UK is in band D or E for both ratings.

6.3.2.2.3.3 EPCs Recommendations The certificate includes recommendations on ways to improve the home's energy efficiency to save money [7]. The accuracy of the recommendations will depend on the inspection standards applied by the inspector, which may be variable. Inspectors, who may be home inspectors (HIs) or domestic energy assessors (DEAs), are audited by their accreditation bodies in order to maintain standards [8]. The recommendations appear general in tone but are in fact bespoke to the property in question. The logic by which the RDSAP programme makes its recommendations was developed as part of a project to create the RDSAP methodology, which took place during the early years of the 21st century [9]. The EU directive requires the EPC recommendations to be cost-effective in improving the energy efficiency of the home but, in addition to presenting the most cost-effective options, more expensive options that are less cost-effective are also presented. To distinguish them from the more cost-effective measures, these are shown in a section described as 'further measures'. Because the EPC is designed to be produced at a change of occupancy, it must be relevant to any occupier; it therefore must make no allowance for the particular preferences of the current occupier.

6.3.2.2.3.4 Exempt Properties Properties exempt from the Housing Act 2004 are

1. Non-residential, such as offices, shops, warehouses.
2. Mixed use—a dwelling house that is part of a business (farm, shop, petrol station).
3. Unsafe properties—a property that poses a serious health and safety risk to occupants or visitors.
4. Properties to be demolished—properties that are due to be demolished, where the marketing of the property, all the relevant documents and planning permission exist.

6.3.2.3 Non-Domestic EPCs

In addition to the requirements in relation to dwellings, there is also a requirement for EPCs in connection with the sale, rent or construction of buildings other than dwellings with a floor area greater than 50 m^2 (from April 2008) that contain fixed services that condition the interior environment.

Properties that are exempt from requiring a domestic EPC will generally require a non-dwelling energy performance certificate, which was also required by the Energy Performance of Buildings Directive. Commercial properties and public buildings currently account for nearly 25% of the UK's carbon emissions, contributing to global climate change [10]. All non-dwelling EPCs must be carried out by, or under the direct supervision of, a trained non-domestic energy assessor, registered with an approved accreditation body. Department for Communities and Local Government (DCLG) has arranged for a publicly accessible central register of such assessors maintained by the Landmark Information Group.

There are three levels of buildings: Level 3, Level 4, and Level 5. The complexity and the services used by that building will determine which level it falls under. They are as follows:

1. Level 3 = small buildings, with heating systems less than 100 kW and cooling systems less than 12 kW

2. Level 4 = purpose built buildings, with heating systems greater than 100 kW and cooling systems greater than 12 kW

3. Level 5 = larger buildings that are complex in shape

A commercial energy assessor must be qualified to the level of the building to carry out the inception.

Beginning in October 2008, all buildings including factories, offices, retail premises and public sector buildings must have an EPC whenever the building is sold, built or rented. Public buildings in England and Wales (but not Scotland) also require a display energy certificate showing actual energy use and not just the theoretical energy rating. Beginning in January 2009, inspections for air-conditioning systems were introduced.

6.3.2.4 Display Energy Certificates

Display energy certificates (DECs) show the actual energy usage of a building and the operational rating, and help the public see the energy efficiency of a building. This is based on the energy consumption of the building as recorded by gas, electricity and other meters. The DEC should be clearly displayed at all times and be clearly visible to the public. A DEC is always accompanied by an advisory report that lists cost-effective measures to improve the energy rating of the building. DECs are only required for buildings with a total useful floor area of more than 1000 m^2 that are occupied by a public authority and institution providing a public service to a large number of persons and therefore visited by those persons. They are valid for one year. The accompanying advisory report is valid for seven years.

However, to make it easier for public authorities with multiple buildings on one site to comply with the legislation, a site-based approach for the first year (to October 2009) is allowed where it is not possible to produce individual DECs. This means that only one DEC will need to be produced based on the total energy consumption of the buildings on the site. Public bodies most affected by this relaxation are NHS Trusts, universities and schools.

The requirement for DECs came into effect in October 2008. They were trialed in the UK under an EU-funded project also called 'Display' and coordinated by Energie-Cités; participants included Durham County Council and the Borough of Milton Keynes.

6.3.2.5 Criticism

EPCs have gained some political controversy, partly reflecting the housing market crisis in the United Kingdom (2008). Many in the housing industry, such as the Royal Institution of Chartered Surveyors, have criticised the introduction of both Home Information Pack (HIPs) and EPCs because they are not automatically qualified to carry out inspections. The EPC suffers deficiencies in attempting to evaluate the energy efficiencies of houses. For example, it takes no account of wall thickness at all, so that some very thick and thus insulating walls in older houses count the same as though they were only two bricks thick. A further objection is often made concerning the quality of inspection made to produce the certificate. It cannot be invasive, so the inspector cannot drill walls or ceilings to determine the state or even existence of any insulation. He can either assume the worst (no insulation present) or rely on the householder (who may know about the matter). This can produce uncertainty about the validity of the output from his or her analysis. In addition, the procedure lacks detail—for example, draught proofing is not considered nor is the balance between ventilation and heat retention for the dwelling. Finally, EPCs pose particular problems for the owners of listed buildings because improvements such as double glazing are often barred by the controls on changes to such structures, making it difficult to rectify low ratings.

6.3.2.6 The Energy Label Australia

The energy rating label enables consumers to compare the energy efficiency of domestic appliances on a fair and equitable basis. It also provides an incentive for manufacturers to improve the energy performance of appliances. The energy rating label was first introduced in 1986 in New South Wales and Victoria. It is now mandatory in all states and territories for refrigerators, freezers, clothes washers, clothes dryers, dishwashers and air conditioners (single phase only) to carry the label when they are offered for sale. Three-phase air conditioners may carry an energy label if the suppliers choose to apply for one.

The energy rating label has two main features:

1. The star rating gives a quick comparative assessment of the model's energy efficiency.
2. The comparative energy consumption (usually kWh/year) provides an estimate of the annual energy consumption of the appliance

based on the tested energy consumption and information about the typical use of the appliance in the home. Air conditioners show the power consumption of the appliance (kW or kWh/h).

The star rating of an appliance is determined from the energy consumption and size of the product. These values are measured under Australian standards that define test procedures for measuring energy consumption and minimum energy performance criteria. Appliances must meet these criteria before they can be granted an energy rating label (Figure 6.13).

Following several years of negotiation between government and industry, the familiar red and yellow energy rating label was revised in 2000. Energy efficiency is now measured against a tougher standard. This change encourages improved technology and more efficient products, which will save consumers money and will help reduce harmful greenhouse gas emissions. An old energy label was in use in Australia from 1986 until 2000 and was being reviewed during 2003 to coordinate and plan future changes across all product groups. The elements under consideration are deciding whether the green band at the base of the label should be retained and making the appliance capacity more prominent on the energy label.

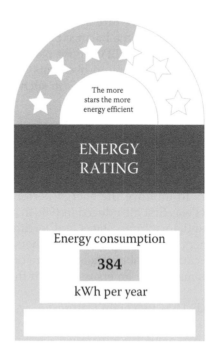

FIGURE 6.13
Example of the energy rating label.

6.3.2.6.1 *Programmes for Other Appliances*

1. Background to the energy labelling programme
2. History of the energy labelling programme
3. Energy labelling: facts and fiction
4. How are star ratings calculated?
5. Tips for purchasing energy-efficient appliances
6. Energy costs: the second price tag
7. Tips on how to save energy when using an appliance
8. Why is energy efficiency important?
9. How are appliances tested for energy consumption?
10. Refrigerators and freezers

6.3.2.6.2 *Refrigerators*
How are Refrigerator Star Ratings Calculated?

6.3.2.6.2.1 Comparative Energy Consumption The energy consumption of a refrigerator is measured under conditions specified in an Australian standard. Over a year, this value is known as the comparative energy consumption or CEC.

6.3.2.6.2.2 Concept of Adjusted Volume The measure of energy service for a refrigerator is the volume that has to be cooled. However, freezers operate at a much colder temperature than fresh food compartments, so the volume of these compartments is 'adjusted' to take into account the extra energy required to achieve this task. The energy service provided by a refrigerator or freezer is called the total adjusted volume.

6.3.2.6.2.3 Performance To be eligible for an energy label, a refrigerator has to meet a temperature operation test and a pull down test as defined in the Australian standard. Refrigerators also have to meet the minimum energy performance standards (MEPS) requirements for minimum energy efficiency.

6.3.2.7 Base Energy Consumption and Star Rating

The base energy consumption (BEC) defines the 'one star' line for particular products. An additional star is awarded when the CEC of the model is reduced by a defined percentage from the BEC. Energy reductions per star from 2010 were set at 23% per star. For example, a CEC that was 0.77 of the BEC or less would achieve two stars. Similarly, a CEC of 0.5929 (0.77×0.77) of the BEC or less would achieve three stars and so on.

For refrigerators and freezers, there are a total of 10 groups that have been defined under the standard. However, to determine star ratings, a number of these groups are compared on the same basis, especially where they are performing a similar task. From 2010, there are only three BEC rating scales which cover all 10 refrigerator and freezer groups.

Refrigerators and freezers are a special category of product under the energy labelling programme because they are also subjected to MEPS. The effect of the MEPS requirements for refrigerators and freezers was considered when the BEC lines were developed. The new BEC lines from 2010 are based on curves that have been developed on the basis of estimated surface (adjusted volume to the power of two-thirds) area rather than adjusted volume, which has been used previously; this helps reduce the effect of size bias.

What about Adaptive Defrost?
Adaptive defrost is smart technology for frost-free refrigerators and freezers where the appliance adjusts the frequency of automatic defrosts to match frost load on the evaporator. Adaptive defrost systems are recognised by the Australian standard for refrigerators and freezers, but at this stage, a realistic test to assess the performance of these controls is not included in the test method. The standard assumes that at least one defrost per day will occur during normal operation, although some smart systems may have a longer time between defrosts under normal conditions of use. An adaptive defrost allowance of 1.05 is included in the MEPS levels, but this allowance is not included in star rating equations.

6.3.2.7.1 Clothes Washers

1. CEC: The energy consumption of a clothes washer is measured under conditions specified in an Australian standard. Over a year, it is assumed that the clothes washer is used seven times per week at rated capacity on a warm wash (warm CEC in red). A value for cold wash energy of seven times per week is also shown on the label (cold CEC in blue). The clothes washer is labelled on the 'normal' or 'regular' programme (programme specified for a normally soiled cotton load). The energy consumption of a clothes washer includes electrical energy for motors and pumps and the energy embodied in any imported hot water or electrical energy used to heat the water internally. The majority of energy for a clothes washer is used to heat water on a warm wash.

2. Capacity: The measure of energy service for a clothes washer is rated load capacity. This is the value declared by the manufacturer and defines the test load used in the Australian standard.

3. Performance: To be eligible for an energy label, a clothes washer must be able to meet a minimum level of wash performance and

a minimum level of spinning performance and must not exceed the 'wear and tear' limits that are defined in the Australian standard.

4. BEC and star rating: The BEC defines the 'one star' line for particular products. An additional star is awarded when the CEC of the model is reduced by a defined percentage from the BEC. The energy reduction per star is 27% for clothes washers. For example, a model that had a CEC that was 0.73 of the BEC or less would achieve two stars. Similarly, a CEC of 0.533 (0.73 × 0.73) of the BEC or less would achieve three stars and so on.

For clothes washers, front and top loading models are rated on the same basis. The warm wash energy consumption and a component of residual moisture (spin performance) are used to define the star rating in comparison with the BEC. Therefore, a model that has a good spin performance may get a marginally higher star rating than a model of the same capacity and CEC with a poor spin performance. The detailed star rating equations are contained in the document 'Equations for Appliance Star Ratings'.

6.3.2.7.2 Clothes Dryers

1. CEC: The energy consumption of clothes dryer is measured under conditions specified in an Australian standard. Over a year, it is assumed that the clothes dryer is used at rated capacity once per week (52 times per year). The initial moisture content of the clothes load is also defined in the standard.

2. Capacity: The measure of energy service for a clothes dryer is rated load capacity. This is the value declared by the manufacturer and defines the test load used in the Australian standard for the determination of energy consumption.

3. Performance: To be eligible for an energy label, a clothes dryer must be able to dry a standard load in a single operation. Other requirements are a maximum clothes temperature limit of less than 130°C (to prevent scorching) and minimum efficiency requirements that are defined in the Australian standard.

4. BEC and star rating: The BEC defines the 'one star' line for particular products. An additional star is awarded when the CEC of the model is reduced by a defined percentage from the BEC. The energy reduction per star is 15% for clothes dryers. For example, a model that had a CEC that was 0.85 of the BEC or less would achieve two stars. Similarly, a CEC of 0.723 (0.85 × 0.85) of the BEC or less would achieve 3 stars and so on.

For clothes dryers, timer and autosensing models are treated slightly differently: timer models are given a 10% penalty on energy (and for the subsequent calculation of the star rating) on the basis that the way timer controls are used in normal practice results in some overdrying of the clothes load. Under the standard test, timer dryers are operated until the load reaches a final moisture content of 6%, although autosensing dryers are operated until they terminate their drying automatically (but at a moisture content of less than or equal to 6%), so the tested difference is usually less than 10%. There is an overview of how star ratings are calculated for other products on this site.

6.3.2.7.3 Dishwashers

1. CEC: The energy consumption of a dishwasher is measured under conditions specified in an Australian standard. Over a year, it is assumed that the dishwasher is used seven times per week (365 times per year). The programme used for the energy labelling programme is currently the one specified by the manufacturer, although by April 2004, all dishwashers were re-labelled on their 'normal' programme using the revised AS/NZS 2007 Part 1 test method released in 2003.

2. Capacity: The measure of energy service for a dishwasher is the number of place settings. This is the value declared by the manufacturer and defines the test load used in the Australian standard.

3. Performance: To be eligible for an energy label, a dishwasher must be able to meet the specified wash and dry performance criteria defined in the Australian standard.

4. BEC and star rating: The BEC defines the 'one star' line for particular products. An additional star is awarded when the CEC of the model is reduced by a defined percentage from the BEC. The energy reduction per star is 30% for dishwashers. For example, a model that had a CEC that was 0.70 of the BEC or less would achieve two stars. Similarly, a CEC of 0.49 (0.70×0.70) of the BEC or less would achieve three stars and so on.

The detailed star rating equations are contained in the document 'Equations for Appliance Star Ratings'. There is an overview of how star ratings are calculated for other products on this site.

6.3.2.7.4 Air Conditioners

1. Power input (also called comparative energy consumption): The energy consumption or power input of an air conditioner is measured under conditions specified in an Australian standard. Because air-conditioner use is affected by climate and this varies substantially across Australia, the cooling and/or heating power input shown on the energy label is the energy the air conditioner uses per hour at rated capacity (the units on the label are in kW, which is the same as kWh/h). To work out the annual energy use will require information on the climate and other factors such as occupancy (hours that

cooling is required) and building shell performance (insulation, glazing, orientation etc.). It is important to note that, under normal usage, the air conditioner will spend a significant amount of time at less than its rated capacity; in terms of efficiency, this is really only important for variable speed drive models.

2. Capacity output: The measure of energy service for an air conditioner is the rated cooling and/or heating capacity of the air conditioner, usually specified in kW. (Some product brochures use BTUs, although this is now unusual. Some retailers use compressor 'horsepower', but this has no meaning in terms of the units' capability.) These rated values are as declared by the manufacturer and the test conditions are defined in the Australian standard. The heating capacity of a reverse cycle air conditioner is the heat that can be put into a room. Similarly, the cooling capacity is the heat that can be removed from a room. The cooling capacity is made up of the sensible component (usually the majority of the capacity), which relates to the actual temperature reduction (cooling) of the air, plus the latent component, which is a measure of the dehumidification effect of the indoor air. Latent cooling capacity is sometimes expressed as moisture removal capacity in litres or kilograms of water per hour (1 kg per hour of moisture removal is equal to 683 W latent capacity).

3. How can the capacity output be greater than the power input? Refrigerative air conditioners (the only type covered by energy labelling in Australia; evaporative units are not included) use a technique called the vapour compression cycle to 'move' energy in the form of heat from one space to another. This is generally a very efficient process, and the amount of heat moved is typically two to three times (or more) the energy required to run the compressor system. This ratio is called the energy efficiency ratio (cooling) or coefficient of performance and is used as the basis for determining the star rating of an air conditioner (see following). The efficiency of the system depends on the components used (their design and how well these are matched) and the temperature difference between inside and outside (as the temperature difference increases, the system becomes less efficient). The system uses a refrigerant (which exists as a gas at low pressure and as a liquid under compression) that is compressed and liquefied, allowed to cool in a condenser, and then allowed to expand to become a gas in an evaporator (the expansion is accompanied by a strong cooling effect). In this operation, the condenser becomes warm and the evaporator becomes cold as the heat is moved from the evaporator to the condenser.

The principle is the same as used in a normal refrigerator, which 'moves' heat from the inside of refrigerator to the outside. In the case of an air conditioner, when in cooling mode, the heat is removed from the room being cooled and pushed outside through the refrigeration system. Similarly, if the unit can operate in 'reverse' (the so-called heating mode or reverse cycle), the process runs backward and the energy is collected from outside and moved inside to the room being heated.

1. Performance: To be eligible for an energy label (and to comply with MEPS), an air conditioner must meet the maximum cooling test as defined in the Australian standard; this ensures that the air conditioner is capable of operating under extreme conditions. The air conditioner also has to have a tested capacity of not less than 95% of the rated value and a tested energy consumption of not more than 105% of the rated value.

2. Star rating: The star rating for air conditioners is determined differently than for other appliances. For air conditioners, the measure of energy efficiency is the EER for cooling and the COP for heating. The EER and COP are defined as the capacity output divided by the power input. The star rating index is calculated on the tested values for energy and capacity, rather than the nameplate or rated values. Typically, the EER and COP are in the range 2.0–3.5 (meaning that the cooling or heating output is 2–3.5 times as great as the power input, or an efficiency of 200%–350%). This is achieved by the use of a refrigeration heat pump that collects internal heat and moves it outside when in cooling mode or collects ambient heat from outside and moves it inside when in heating mode. The apparent efficiency of heat pumps is high because they can move much lower grade energy in the form of heat than they require as electrical power input. The star rating for air conditioners is determined from the tested EER and COP. For cooling, one star is equal to an EER of 2.0 with an extra star for an increase in EER of 0.3. For heating, one star is equal to a COP of 2.3 with an extra star for an increase in COP of 0.3. From 2010, one star was equal to an EER and COP of 2.75 with a step of 0.5 for each additional star. Importantly, from 2010, the star rating will be based on an annual efficiency calculation that includes any non-operational energy consumption.

What Do the Labels Look Like for Each of the Labelled Products?

1. Requirements for gas space heaters, gas water heaters and gas cookers: Energy labels can be found on gas space heaters (AS4553 and AS4556) and gas water heaters (storage and instantaneous) (AS4552). Gas energy labels are similar in format to those found on electrical appliances, except they are blue in colour and annual energy is shown in megajoule. The gas labelling programme is currently an industry voluntary scheme that was once managed by the Australian Gas Association (AGA).

Energy tests, label requirements and relevant performance requirements are specified in following standards:

a. AS 4551 (AG 101) Domestic Gas Cooking Appliances

b. AS 4552 (AG 102) Gas Water Heaters

c. AS 4553 (AG 103) Gas Space Heating Appliances

d. AS 4554 (AG 104) Gas Laundry Dryers

e. AS 4555 (AG 105) Domestic Gas Refrigerators

f. AS 4556 (AG 106) Indirect Gas-Fired Ducted Air Heaters

g. AS 4557 (AG 107) Domestic Outdoor Gas Barbecues

h. AS 4558 (AG 108) Decorative Gas Log Fires and Other Fuel Effect Appliances (Figures 6.14 through 6.19).

2. Strategy and work programme for gas appliances: The main objective of the Switch on Gas ten-year strategic plan is to implement a nationally consistent regulation scheme for energy efficiency of gas appliances and equipment. This strategy is an important part of the package of measures being implemented by the MCE under the National Framework for Energy Efficiency (NFEE).

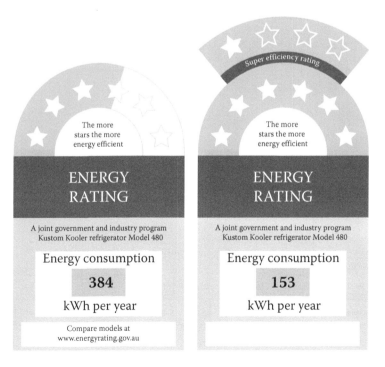

FIGURE 6.14
Sample of the Australian energy labels for refrigerators and freezers AS/NZS 4474.2:2009.

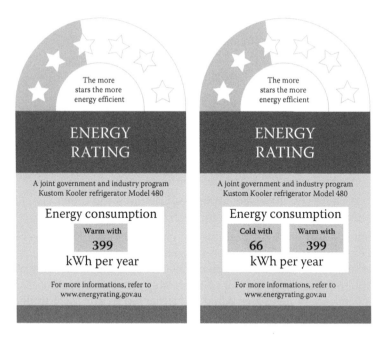

FIGURE 6.15
Sample of the Australian energy labels for clothes washers AS/NZS 2040.2:2005.

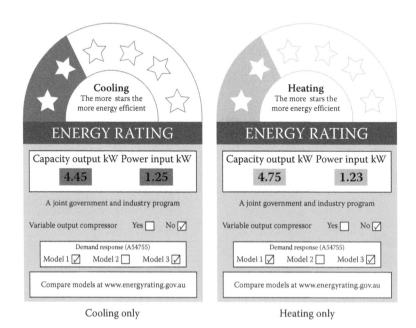

FIGURE 6.16
Sample of the Australian energy labels for room air conditioners AS/NZS 3823.2:2009.

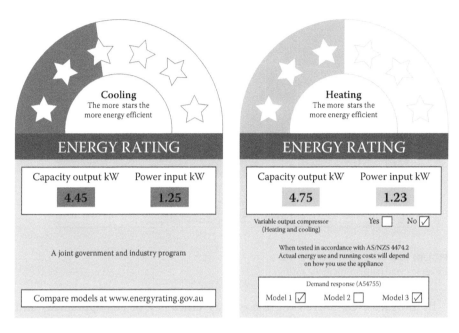

FIGURE 6.17
Cooling only, heating only and reverse cycle labels.

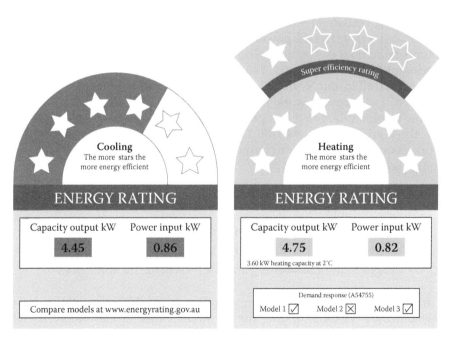

FIGURE 6.18
Energy labels for super efficiency units.

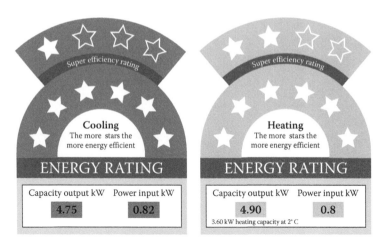

FIGURE 6.19
Reverse cycle (super efficiency rating).

By progressively increasing the energy efficiency of gas appliances and equipment beyond business as usual levels, Switch on Gas responds to the evolving needs of the community by providing world class energy-efficient gas products. Switch on Gas will also make a contribution to national efforts to reduce greenhouse gas emissions. Natural gas currently supplies about 30% of total household energy in Australia. Within 20 years it is projected that Switch on Gas has the potential to reduce Australian consumers' expenditures on natural gas by up to $115 million per annum and to reduce their consumption by more than 5% below business as usual.

The Australian and New Zealand Equipment Energy Efficiency Program released its inaugural work plan for Switch on Gas in April 2005, outlining products targeted for potential regulation in both countries for the three-year period, 2005/06 to 2007/08. An updated version of this work plan was released for comment in October 2006 (2006/12) and is a consequence of

1. The administrative and regulatory basis of the gas appliance efficiency programme
2. New information about gas appliance technology and testing
3. Gas consumer response to energy labels
4. Changes to the gas labelling and MEPS scheme for appliances

In 2002, the Sustainable Energy Authority of Victoria (SEAV) (now Sustainability Victoria, or SV), the Victorian Office of Gas Safety (OGS), and the Australian Greenhouse Office (AGO) commenced working together to review the gas appliance labelling and MEPS scheme and to explore options for enhancing its effectiveness at driving energy efficiency improvements. The first step in this process was to commission a consultant study on the gas appliance scheme.

A joint government–industry working group comprising SEAV, OGS, AGO, AGA and the GAMAA agreed to work cooperatively to enhance the effectiveness of the gas appliance efficiency scheme. A discussion paper developed on behalf of the joint working group sets out a range of options for the future direction of the gas appliance efficiency scheme. The aim of the discussion paper 'Driving Energy Efficiency Improvements to Domestic Gas Appliances' is to stimulate debate on the future directions of the gas appliance efficiency programme and to canvas views from a broad spectrum of industry and other stakeholders. In 2006 and 2007, work has been undertaken on the development of a revised energy test method for gas water heaters which is expected to be released for public comment in late 2007 of early 2008. An example of the gas appliance energy label is shown in Figure 6.20.

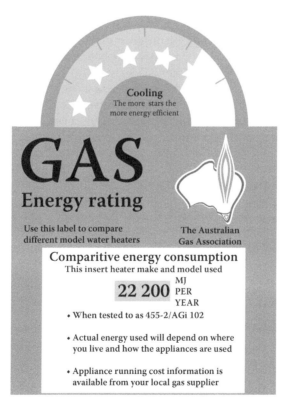

FIGURE 6.20
Gas appliance label.

6.3.2.8 U.S. Energy Star

Energy Star is an international standard for energy-efficient consumer products that originated in the United States. It was first created as a U.S. government programme by the Clinton administration in 1992, but Australia, Canada, Japan, New Zealand, Taiwan, and the EU have also adopted the programme. Devices carrying the Energy Star logo, such as computer products and peripherals, kitchen appliances, buildings and other products, generally use 20%–30% less energy than required by federal standards. However, many European targeted products are labelled using a different standard, a combined energy usage and ergonomics rating from the Swedish Confederation of Professional Employees (TCO) instead of Energy Star.

6.3.2.8.1 History

The Energy Star programme was created in 1992 by the U.S. EPA in an attempt to reduce energy consumption and greenhouse gas emission by power plants; the programme's goal was to demonstrate the potential for profit in reducing greenhouse gases and to facilitate further steps in reducing global warming gases.

Initiated as a voluntary labelling programme designed to identify and promote energy-efficient products, Energy Star began with labels for computer products. In 1995, the programme was significantly expanded, introducing labels for residential heating and cooling systems and for new homes. As of 2012, more than 65,000 Energy Star products were available in a wide range of items including major appliances, office equipment, lighting, home electronics and more. In addition, the label can also be found on new homes and on commercial and industrial buildings. In 2012, about 25% of new housing in the United States was labelled Energy Star.

The EPA estimates that it saved about $20 billion in energy costs in 2012 alone. The Energy Star programme has helped spread the use of light-emitting diode (LED) traffic lights, efficient fluorescent lighting, power management systems for office equipment and low standby energy use.

In 2008, the EPA announced the Green Power Partnership programme, which was designed to help achieve its goal of encouraging the use of renewable power sources. The renewable energy credits allow companies without direct access to renewable power to achieve their goals. However, to avoid companies buying Renewable Energy Centers (RECs) years in advance of any of the hypothetical power ever being produced, RECs are only accepted into the programme when the actual equivalent renewable power will be produced.

6.3.2.8.2 Specifications

Energy Star specifications differ with each item and are set by either the EPA or the U.S. DOE. The following sections highlight product and specification information available on the Energy Star website.

6.3.2.8.3 Computers

New Energy Star 4.0 specifications for computers became effective on 20 July 2007. They require the use of 80 PLUS bronze level or higher power supplies. Energy Star 5.0 became effective on 1 July 2009.

6.3.2.8.4 Servers

The EPA released Version 1.0 of the Computer Server specifications on 15 May 2009. It covers stand-alone servers with one to four processor sockets. A second tier to the specification covering servers with more than four processor sockets, as well as blade servers and fault-tolerant machines was revised in 2011.

6.3.2.8.5 Appliances

Recently, average refrigerators need 20% savings over the minimum standard. Dishwashers need at least 41% savings. Most appliances as well as heating and cooling systems have a yellow energy guide label showing the annual cost of operation compared to other models. This label is created through the DOE and often shows if an appliance is Energy Star [7]. Although an Energy Star label indicates that the appliance is more energy efficient than the minimum guidelines, purchasing an Energy Star labelled product does not always mean you are getting the most energy-efficient option available. For example, dehumidifiers that are rated under 25 US pints (12 L) per day of water extraction receive an Energy Star rating if they have an energy factor of 1.2 (higher is better), while those rated 25 US pints (12 L) to 35 US pints (17 L) per day receive an Energy Star rating for an energy factor of 1.4 or higher. Thus a higher capacity but non–Energy Star rated dehumidifier may be a more energy-efficient alternative than an Energy Star rated but lower capacity model [8]. The Energy Star programme's savings calculator has also been criticised for unrealistic assumptions in its model that tend to magnify savings benefits to the average consumer [9].

Another factor yet to be considered by the EPA and DOE is the overall effect of energy-saving requirements on the durability and expected service life of a mass market appliance built to a consumer-level cost standard. For example, a refrigerator may be made more efficient by the use of more insulative spacing and a smaller capacity compressor using electronics to control operation and temperature. However, this may come at the cost of reduced interior storage (or increased exterior mass) or a reduced service life due to compressor and/or electronic failures. In particular, electronic controls used on new generation appliances are subject to damage from shock, vibration, moisture, or power spikes on the electrical circuit to which they are attached. Critics have pointed out that even if a new appliance is energy efficient, any consumer appliance that does not provide customer satisfaction or that must be replaced twice as often as its predecessor contributes to landfill pollution and wastage of natural resources used to construct its replacement [10].

6.3.2.8.6 Heating and Cooling Systems

Energy Star qualified heat pumps, boilers, air-conditioning systems and furnaces are available. In addition, cooling and heating bills can be significantly lowered with air sealing and duct sealing. Air sealing reduces the outdoor air that penetrates a building, and duct sealing prevents attic or basement air from entering ducts and lessening the heating/cooling system's efficiency. Energy Star–qualified room air conditioners are at least 10% more energy efficient than the minimum U.S. federal government standards [11].

6.3.2.8.7 Home Electronics

Energy Star–qualified televisions [12] use 30% less energy than average. In November 2008, television specifications were improved to limit on-mode power use in addition to standby power, which is limited by the current specifications. A wider range of Energy Star qualified televisions will be available. Other qualified home electronics include cordless phones, battery chargers, and external power adapters, most of which use 90% less energy.

6.3.2.8.8 Lighting

The Energy Star is awarded to only certain bulbs that meet strict efficiency, quality and lifetime criteria. Energy Star qualified fluorescent lighting uses 75% less energy and lasts up to ten times longer than normal incandescent lights.

Energy Star Qualified LED Lighting:

1. Reduces energy costs: uses at least 75% less energy than incandescent lighting, saving on operating expenses.

2. Reduces maintenance costs: lasts 35–50 times longer than incandescent lighting and about two to five times longer than fluorescent lighting—no bulb replacements, no ladders, no on-going disposal programme.

3. Reduces cooling costs: LEDs produce very little heat.

4. Is guaranteed: comes with a minimum three-year warranty, far beyond the industry standard.

5. Offers convenient features: available with dimming on some indoor models and automatic daylight shut-off and motion sensors on some outdoor models.

6. Is durable: will not break like a bulb.

To qualify for Energy Star certification, LED lighting products must pass a variety of tests to prove that the products will display the following characteristics:

1. Brightness is equal to or greater than existing lighting technologies (incandescent or fluorescent), and light is well distributed over the area lighted by the fixture.
2. Light output remains constant over time, only decreasing towards the end of the rated lifetime (at least 35,000 hours or 12 years, based on use of eight hours per day).
3. Excellent colour quality; the shade of white light appears clear and consistent over time.
4. Efficiency is as good as or better than fluorescent lighting.
5. Light comes on instantly when turned on.
6. No flicker when dimmed.
7. No off-state power draw: the fixture does not use power when it is turned off, with the exception of external controls, whose power should not exceed 0.5 W in the off state.

6.3.2.8.9 Desktop Computers

A new Energy Star specification for desktop computers went into effect in July 2007 [13]. The requirements are more stringent than the previous specification, and existing equipment designs can no longer use the logo unless re-qualified. The power requirements are for 80% or greater AC power supply efficiency using the standards defined by the 80 Plus Program [14].

6.3.2.8.10 New Homes

New homes that meet strict guidelines for energy efficiency can qualify for Energy Star certification. An Energy Star qualified home uses at least 15% less energy than standard homes built to the 2004 International Residential Code (IRC). They usually include properly installed insulation, high performance windows, tight construction and ducts, energy-efficient cooling and heating systems, and Energy Star–qualified appliances, lighting and water heaters [15].

6.3.2.9 Energy Performance Ratings

The EPA's Energy Star programme has developed energy performance rating systems for several commercial and institutional building types and manufacturing facilities. These ratings, on a scale of 1–100, provide a means for benchmarking the energy efficiency of specific buildings and industrial plants against the energy performance of similar facilities. The ratings are used by building and energy managers to evaluate the energy performance of existing buildings and industrial plants. The rating systems are also used by the EPA to determine if a building or plant can qualify to earn Energy Star recognition [16].

For many types of commercial buildings, you can enter energy information into EPA's free online tool, Portfolio Manager [2], and it will calculate a score for your building on a scale of 1–100. Buildings that score a 75 or greater may qualify for the Energy Star. Portfolio Manager is an interactive energy management tool that allows you to track and assess energy and water consumption across your entire portfolio of buildings in a secure online environment. Whether you own, manage or hold properties for investment, Portfolio Manager can help you set investment priorities, identify under-performing buildings, verify efficiency improvements and receive EPA recognition for superior energy performance [17].

6.3.2.9.1 *Buildings*

The number of space types that can receive the energy performance rating in Portfolio Manager is expanding and now includes [18]: banks/financial institutions, courthouses, hospitals (acute care and children's), hotels and motels, house of worship, K-12 schools, medical offices, offices, residence halls/dormitories, retail stores, supermarkets, and warehouses (refrigerated and non-refrigerated) [19]. Technical descriptions for models used in the rating system are available for review [3]. These documents provide detailed information on the methodologies used to create the energy performance ratings, including details on rating objectives, regression techniques and the steps applied to compute a rating. Energy Star energy performance ratings have been incorporated into some green buildings standards, such as Leadership in Energy and Environmental Design for existing buildings.

6.3.2.9.2 *Industrial Facilities*

Energy performance ratings have been released for the following industrial facilities [20–25]:

1. Automobile assembly plants, cement plants, wet corn mills, container glass manufacturing, flat glass manufacturing, frozen fried potato processing plants, juice processing petroleum refineries, and pharmaceutical manufacturing plants [19].

6.3.3 Summary of Energy Standards and Labelling

Table 6.3 summarises the international energy labelling activities in recent years that had achieved their goals and their impact on the national economy.

TABLE 6.3

Summary of Pioneering International Programmes and Their Achievements

Country or Region	Programme	Achievements
Australia	Mandatory Standards and Labelling	• 11% reduction in energy consumption of labelled appliances in 1992 • Approximately equals 94 GWh of saved energy or 1.6% decrease in total household electricity consumption
Europe	Mandatory Standards and Labelling	• Germany: 16.1% increase in market efficiency (1993–1996) • Netherlands: 12.6% increase in market efficiency (1992–1995) • United Kingdom: 7.3% increase in refrigerator/freezer efficiency (1994–1996)
Philippines	Mandatory Standards and Labelling	• 25% increase in average efficiency of all air conditioners (after first year) • Energy savings: 6 MW in demand and 17 GWh in consumption (after first year)
Egypt	Mandatory Standards and Labelling	• 10% decrease in refrigerator energy consumption (after three years) • 20% decrease in air-conditioner energy consumption (after three years)
Thailand	Voluntary Labelling	• 14% decrease in refrigerator energy consumption (after three years) • Energy savings: 65 MW in demand and 643 GWh in consumption
United States	Mandatory Standards and Labelling	• 98% increase in refrigerator efficiency (1972–1988) • More than 3% reduction in U.S. annual residential consumption from appliances and lighting equipment

Reference Standards

	Entry into Force – Date of Expiry	Deadline for Transposition in the Member States	Official Journal
Act			
Directive 92/75/EEC	2.10.1992	1.7.1993	OJ L 297 of 13.10.1992
Amending Act(s)			
Regulation (EC) No 1882/2003	20.11.2003	–	OJ L 284 of 31.10.2003
Regulation (EC) No 1137/2008	11.12.2008	–	OJ L 311 of 21.11.2008

Related Acts

Proposal for a Directive of the European Parliament and of the Council of 13 November 2008 on the indication by labeling and standard product information of the consumption of energy and other resources by energy-related products [COM(2008) 778 final—Non-published in the Official Journal].

The Commission proposes a recast of Directive 92/75/EEC and an extension of its scope, currently restricted to household appliances, to all energy-related products, excluding means of transport. The Commission defines energy-related products as those having an impact on energy consumption during use. The Commission also proposes that products which do not comply with minimum requirements in terms of energy efficiency should not be eligible for public procurement or incentives. The recast of the Energy Labeling Directive is among the priorities of the Sustainable Consumption and Production and Sustainable Industrial Policy Action Plan presented by the Commission in June 2008.

Codecision Procedure (COD/2008/0222)

Implementing rules

1. Commission Directive 2003/66/EC (energy labeling of household electric refrigerators, freezers, and their combinations) [Official Journal L 170 of 9.7.2003].

2. Commission Directive 2002/40/EC (energy labeling of household electric ovens) [Official Journal L 128 of 15.5.2002].

3. Commission Directive 2002/31/EC (energy labeling of household air conditioners) [Official Journal L 86 of 3.4.2002].

4. Commission Directive 1999/9/EC (energy labeling of household dishwashers) [Official Journal L 56 of 4.3.1999].

5. Commission Directive 98/11/EC (energy labeling of household lamps) [Official Journal L 71 of 10.3.1998].

6. Commission Directive 96/60/EC (energy labeling of household combined washer dryers) [Official Journal L 266 of 18.10.1996].

7. Commission Directive 95/13/EC (energy labeling of household electric tumble dryers) [Official Journal L 136 of 21.6.1995].

8. Commission Directive 95/12/EC (energy labeling of household washing machines) [Official Journal L 136 of 21.6.1995].

9. Amended by Directive 96/89/EC—[Official Journal L 388, 28.12.1996].

Energy Efficiency

1. Directive 2005/32/EC of the European Parliament and of the Council of 6 July 2005 establishing a framework for the settings of ecodesign requirements for energy-using products and amending Council Directive 92/42/EEC and Directives 96/57/EC and 2000/55/EC of the European Parliament and of the Council [Official Journal L 191 of 22.7.2005].

2. Regulation (EC) No 2422/2001 of the European Parliament and of the Council of 6 November 2001 on a Community energy efficiency labeling programme for office equipment [Official Journal L 332 of 15.12.2001].

3. Communications in the framework of the implementation.

4. Commission communication (energy labeling of household air conditioners—Publication of titles and references of harmonized standards under the Directive) [Official Journal C 115 of 30.4.2004].

5. Commission communication (energy labeling of household combined washer-driers) [Official Journal C 161 of 28.5.1997].

6. Commission communication (energy labeling of household electric refrigerators, freezers and their combinations) [Official Journal C 065 of 1.3.1997].

7. Commission communication (energy labeling of household washing machines) [Official Journal C 312 of 23.11.1995].

8. Commission communication (energy labeling of household electric tumble driers) [Official Journal C 312 of 23.11.1995].

References

1. Tugend, A. (2008). If your appliances are avocado, they're probably not green, *New York Times*, Retrieved June 29, 2008, from http://www.nytimes.com/2008/05/10/business/yourmoney/10shortcuts.html?scp=1&sq=appliances%20avocado%20green&st=cse

2. EnergyStar.gov. (2007). *Milestones: ENERGY STAR*, Retrieved March 1, 2008, from www.energystar.gov/.

3. U.S. Environmental Protection Agency. *2006 Annual Report: Energy Star and Other Climate Protection Partnerships*, Retrieved March 1, 2008, from www.EPA.gov/.

4. EnergyStar.gov. *History: ENERGY STAR*. Retrieved March 1, 2008, from www.energystar.gov.

5. Timmer, J. (2008). *EPA tightens rules on its Green Power Partners*, Arstechnica.com, Retrieved March 23, 2009, from http://arstechnica.com/old/content/2008/12/epa-tightens-rules-on-its-green-power-partners.ars

6. Ng, J. (2009). *New Energy Star 5.0 Specs for Computers Become Effective Today*, DailyTech, Retrieved July 1, 2009, from http://www.dailytech.com/New+Energy+Star+50+Specs+for+Computers+Become+Effective+Today/article15559.htm

7. EnergyStar.gov. *Learn More about Energy Guide: Energy Star*, Retrieved March 1, 2008, from www.energystar.gov/.

8. Green Energy Efficient Homes, Energy Efficient Dehumidifiers, www.solar-aid.org/Green-Energy, 2010.

9. Belzer, R. (2008). Energy Star appliances: EPA's savings calculator exaggerates savings, *Regulatory Economics*, Retrieved March 5, 2008, from neutralsource.org/archives/595.

10. Muñoz, S. S. (2007). Do 'green' appliances live up to their billing, *The Wall Street Journal, Business*, pp. 46–50.

11. EnergyStar.gov. (2008). *Room Air Conditioners Key Product Criteria*, Energystar.gov, Retrieved March 23, 2009, from http://www.energystar.gov/index.cfm?c=roomac.pr_crit_room_ac

12. *California Sustainability Alliance Energy Star Televisions*, Retrieved July 24, 2010, from www.sustainca.org.

13. PowerPulse.net, *New Energy Star Promoting New Specs at APEC and PPDC*, Retrieved March 20, 2006, from PowerPulse.net.

14. 80plus.org. (2007). *The 80 Plus Program | About*, Retrieved March 3, 2007, from https://neea.org/docs/default-source/.../neea-success-story-80-plus.pdf.

15. *ENERGYSTAR Qualified Homes: ENERGYSTAR*. (2009). Energystar.gov, Retrieved March 23, 2009, from http://www.energystar.gov/homes

16. Energystar.gov. (2010). Performance Evaluation. Retrieved March 2010.

17. Energystar.gov. (2011). Retrieved June 2011, from http://www.energystar.gov/index.cfm?c=evaluate_performance.bus_portfoliomanager

18. *Criteria for Rating Building Energy Performance*, Energystar.gov, Retrieved March 23, 2009, http://www.energystar.gov/index.cfm?c=eligibility.bus_portfoliomanager_eligibility

19. Energystar.gov. (2011). Retrieved April 2011, from www.energystar.gov. http://www.energystar.gov/index.cfm?c = business.bus_bldgs

20. *Industries in Focus: ENERGYSTAR*. (2009). Energystar.gov, Retrieved March 23, 2009, from http://www.energystar.gov/index.cfm?c=in_focus.bus_industries_focus#plant

21. Environmental News Service. (2008). *Energy Star Climate Change Claims Misleading, Audit Finds*, Environmental News Service, Washington, DC.

22. *Energy Stars May Not be All They Say They Are*, Housingzone.com, Retrieved March 23, 2009, from http://www.housingzone.com/articleXml/LN888056763.html

23. Why Obama's Energy Savings Estimate May Be Skewed, from www.nytimes.com/2009/02/07/.../07energy.htm

24. Hruska, J. (2009). *Sony LCD Exceeds Energy Star Power Draw 75% of Time*, Arstechnica.com, Retrieved March 23, 2009, http://arstechnica.com/gadgets/news/2009/02/sony-lcd-exceeds-energy-star-power-draw-75-of-time.ars

25. Hruska, J. (2010). *Fake Products and Companies Certified by Energy Star*, Popular Mechanics, Retrieved March 26, 2010, http://www.popularmechanics.com/technology/industry/4350335.html

7

Low Carbon Buildings

7.1 Energy-Efficient Designs of Low Carbon Buildings

7.1.1 Summary

Developing communities on their path toward rapid development are endeavouring to make all necessary and appropriate measures to enhance the efficiency of energy utilisation and to increase the beneficiation of the energy resources. Energy production, transmission, distribution and utilisation efficiency become vital factors and measures of national development. Governmental organisations have been established that are responsible energy planning and efficient utilisation, information dissemination and capacity building as well as for devising the necessary codes and standards. Throughout the nation, energy resources are widely used and consumption rates are, in general, exceeding the internationally accepted values. Energy rationalisation and audit exercises have been developed and monitored by governmental authorities, universities and research centres over the past two decades with definitive positive energy reduction and beneficiation. The development of the relevant codes for residential and commercial energy efficiency in buildings is underway through the governmental bodies responsible for research and development in the building technology sector; it is under this umbrella that the National and Unified Arab Codes are developed and issued. A proposed new Energy Performance in Buildings Directive would fulfil the following main targets of the Energy Performance Directive:

> Legislative authorities shall ensure that, when buildings are constructed, sold or rented out, an energy performance certificate is made available to the owner or by the owner to the prospective buyer or tenant, as the case might be. ...
>
> The energy performance certificate for buildings shall include reference values such as currant legal standards and benchmarks in order to make it possible for consumers to compare and assess the energy performance of the building. The certificate shall be accompanied by recommendations for cost-effective improvement of the energy performance... [Available at www.epbd.eu]

The following steps are required for the energy certification:

1. Develop methodologies for energy declaration of the buildings
2. Develop reference values (key numbers) and/or systems for benchmarking
3. Provide a labelling system for selected buildings
4. Describe an energy signature for the building

As the world becomes increasingly dependent on electrical appliances and equipment, energy consumption rapidly rises every year. Many programmes have been established in various countries to increase end-use equipment energy efficiency. One of the most cost-effective and proven methods for increasing energy efficiency of electrical appliances and equipment is to establish energy efficiency standards and labels. Energy efficiency standards are a set of procedures and regulations that prescribe the energy performance of manufactured products, sometimes prohibiting the sale of products less energy efficient than the minimum standard. The term 'standard' commonly encompasses two possible meanings:

1. A well-defined protocol (or laboratory test procedure) by which to obtain a sufficiently accurate estimate of the energy performance of a product in the way it is typically used or at least a relative ranking of the energy performance compared to other models
2. A target limit on energy performance (usually a maximum use or minimum efficiency) formally established by a government-based agency upon a specified test standard

Energy efficiency labels are informative labels affixed to manufactured products indicating a product's energy performance (usually in the form of energy use, efficiency and/or cost) in order to provide consumers with the data necessary for making informed purchases. Energy labels serve as a complement to energy standards. They provide consumers with information that allows those who care to be able to select more efficient models. Labels also allow utility companies and government energy conservation agencies to offer incentives to consumers to buy the most energy-efficient products. The effectiveness of energy labels is highly dependent on how information is presented to the consumer.

7.1.2 Rationale and Benefits

Energy efficiency in developed and developing countries plays an important role in achieving global sustainable development. Energy consumption is growing rapidly in these countries, yet energy efficiency remains far below levels in developed countries. Energy efficiency improvements can slow the growth in energy consumption, save consumers money and can reduce capital expenses for energy infrastructure [1–11]. For most developing countries, the foreign exchange needed to finance energy sector expansion is a significant drain on reserves. Additionally, energy efficiency reduces local environmental impacts, such as water and air pollution from power plants, and mitigates greenhouse gas emissions. Standards and labelling programmes provide enormous energy-saving potential that can direct developing countries toward sustainable energy use. Improved end-use efficiency from standards and labelling programmes can contribute significantly to developing economies. The main benefits are

1. Less need to build new power plants: The cost of saving 1 kWh of energy through energy efficiency programmes has proven much less expensive than producing 1 kWh of energy by building a new power plant.
2. Reduced greenhouse gas emissions: Less energy production means less carbon dioxide emissions from power plants. This contributes to environmental benefits such as slowing down environmental pollution and global warming and preserving natural resources and the ecosystem.
3. Improved competitiveness for local manufacturers: Local companies that upgrade the efficiency of their products can compete better with multinational companies, especially with lower production costs.
4. Higher consumer disposable income: Lower spending on electric bills increases consumer purchasing power for other products, which helps local businesses.
5. Increased cash flow in the local economy: With higher disposable income, consumers are more willing to spend, thus injecting money into the local economy.
6. Improved trade balance: Decrease in energy demand will reduce the consumption of indigenous resources (i.e. natural gas and oil), allowing more to be exported (for Lebanon, less to be imported). Increased export earnings (or less import spending) help alleviate trade deficits of Arabian countries.
7. Avoid future energy deficit as power demand rises: Energy exporting countries have become net importers due to dramatic increases in electricity demand. Energy efficiency programmes can help slow down the demand and prevent energy deficit in the future.

7.1.3 The Holistic Approach: Think 'Pyramids'

The assessment of the overall energy performance of a building, including the technical building systems, comprises a number of successive steps, which can be schematically visualised as a pyramid (Figure 7.1) [11].

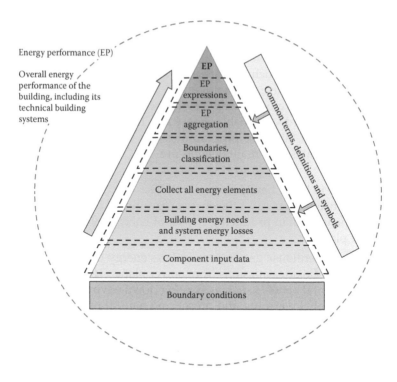

FIGURE 7.1
Overall building energy performance. (From Van Dijk, D. and Khalil, E. E., *ISO Focus.*, 25–27, 2011.)

Sets of common terms, definitions and symbols are essential for all segments from top to bottom. These include energy needs, technical building systems, auxiliary energy use, recoverable system losses, primary energy and renewable energy (Figure 7.2) [11].

The top segment of the pyramid is the main output: the energy performance and the energy performance certificate of the building.

The second segment provides the inputs for the top segment: one or more numerical indicators expressing the energy performance (such as overall energy use per square meter conditioned floor area, EP), a classification and ways to express the minimum energy performance requirements (EP_{max}) (Figure 7.3) [9].

The third segment describes the principles and procedures on the weighting of different energy carriers (such as electricity, gas, oil or wood) when they are aggregated to overall amount of delivered (and exported) energy. For instance, this may be expressed as total primary energy (E_P) or carbon dioxide emission (E_{CO_2}).

The fourth segment specifies the categorisation of building types (for e.g. office spaces, residential or retail) and specification of the boundaries of the building (Figure 7.4) [11].

The fifth segment provides procedures on the breakdown of the building energy needs and system energy losses, aiming to gain clear insights into where energy is used.

The sixth segment provides the building energy needs and energy use for each application (heating, cooling etc.) and the interactions between them.

The seventh segment provides the input data on components, such as thermal transmission properties, air infiltration, solar properties of windows and energy performance of system components and efficiency of lighting.

FIGURE 7.2
Harmonisation of terms is essential. (From Van Dijk, D. and Khalil, E. E., *ISO Focus.*, 25–27, 2011.)

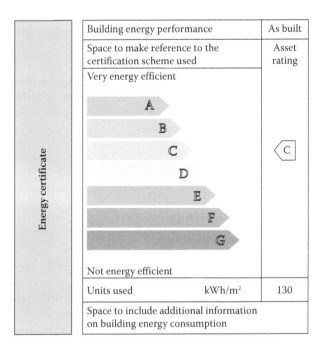

Building energy performance		As built
Space to make reference to the certification scheme used		Asset rating
Very energy efficient		
A		
B		
C		◁ C
D		
E		
F		
G		
Not energy efficient		
Units used	kWh/m²	130
Space to include additional information on building energy consumption		

Energy certificate

FIGURE 7.3
Example of energy performance certificate. (From ISO publications ISO 23045_2009, January 2009.)

ISO/TC 163/WG 4 (Joint TC 163 – TC 205 WG)

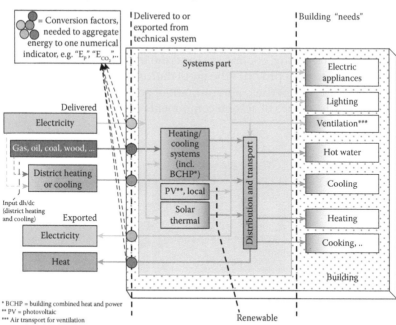

* BCHP = building combined heat and power
** PV = photovoltaic
*** Air transport for ventilation

FIGURE 7.4
Boundaries and main building energy performance elements. (From Van Dijk, D. and Khalil, E. E., *ISO Focus.*, 25–27, 2011.)

The standards that provide the procedures on boundary conditions comprise external climatic conditions, indoor environment conditions (thermal and visual comfort, indoor air quality etc.), standard operating assumptions (occupation) and national legal restrictions.

7.2 Energy-Efficient Buildings

7.2.1 Energy Declaration of Buildings

The primary use of energy declarations is

1. To create consciousness of energy efficiency in buildings and also improve the knowledge of energy use in buildings
2. To determine if a building works as well as possible with regard to its technical design
3. For benchmarking

4. For suggesting measures and recommendations for reducing the energy use

5. To provide the information necessary to make calculations of the environmental impact due to the energy use (e.g. CO_2 emissions)

6. To describe selected energy properties of the building

7. Give the basis for a common energy performance certification of a building

Depending on the purpose of the energy declaration, different procedures can be of interest. Different actors need different information. For example, for benchmarking and for explaining the CO_2 emissions, it can be enough to read the total energy supply to the building and only adjust these figures to normal outdoor climate and to the heated area. For giving relevant advice to the property owner regarding cost-effective measures, a very careful examination and calculation of the building's energy balance is necessary.

One way to proceed is to make the energy calculation in different steps for existing buildings. The first is to collect measured energy use (e.g. from energy bills) and create a benchmark to decide if the actual building is better or worse compared to similar buildings. If the energy use seems to be higher than the average for a comparable grouping of buildings, a second step is to make a careful energy calculation that can be compared to the measured energy use. This has to be done to identify what measures can be recommended in order to reduce the energy use in the building. For benchmarking, it might also be of interest to compare the measured energy use in the building examined with the estimated energy use in a building that is built with the best available technology. Alternatively, it could be compared with a building that meets the requirement in the existing building codes. Some important aspects necessary to take into consideration when developing a common tool for energy declaration of buildings are discussed in the next section. The discussion here is focused on residential buildings, but similar principles are relevant for other types of buildings.

7.2.2 Energy Declaration of Existing Buildings

For most existing buildings, the energy use usually is well known via the energy bills. On the other hand, the construction details usually are not very well documented. Calculations of the energy use will thus in most cases be very uncertain, and difficulties will occur when giving relevant recommendations of cost-effective energy conservation measures.

The energy declaration needs a combination of tools for calculations based on the information from the energy bills. For existing buildings, the energy declarations can be based on measurements and calculations in order to fulfil the purposes mentioned earlier. Normally, the heating bills are based on measured heat delivered to the building. In most cases, the energy for domestic hot water is included in the heating bill. To get the total energy use in a building, the electricity to run the building and household electricity have to be added. In electrically heated houses (common in, e.g. Sweden), the homeowner just has one bill covering the total energy use. To make the measured values objective and comparable, several corrections and calculations are necessary:

1. Verify that the indoor thermal comfort and air quality meet agreed requirements
2. Make corrections for the heat use to normal outdoor climate—primarily outdoor temperature (maybe also solar heat gain)
3. Make necessary corrections for internal heat gains (e.g. differences in household electricity)

7.2.3 Energy Declaration of New Buildings

The energy declaration of a new building has to be based on calculations. In comparison with existing buildings, details of the building construction are available. The most critical part for the outcome of calculations is the choice of input data. To achieve good comparability, a common procedure to determine input data (such as energy for domestic hot water, household electricity or electricity for the activity in the building, choice of indoor temperature, electricity for operating the building, energy for lighting etc.) has to be developed.

In many countries, numerous new buildings have very low energy demand for heating. The solar heat gain, internal gains and energy losses from equipment and so forth cover the major part of the heating demand. In several buildings, the internal gains are so large that cooling is necessary even in temperate climate. For buildings with large glassed areas and/or large internal gains, the energy calculations need to be done on an hourly basis. Many modern apartment blocks, offices, education buildings and restaurants need very little heat supply from the heating system but very often need air-conditioning to attain acceptable thermal comfort.

7.2.4 Issues for International Collaboration

1. Develop standardised tools for the calculation of the energy performance of buildings taking into account the factors outlined in ISO 13790 that cover many aspects but still have to be completed.
2. Define system boundaries for the different building categories and different heating systems.

3. Prepare models for expressing requirements on indoor air quality, thermal comfort in winter and when appropriate in summer, visual comfort and so on.

4. Develop transparent systems to determine necessary input data for the calculations, including default values on internal gains.

5. Provide transparent information regarding output data (reference values, benchmarks etc.)

6. Define comparable energy-related key values (kWh/m^2, kWh per person, kWh per apartment, kWh per produced unit etc.) The areas/volumes need to be defined.

7. Develop a method to translate net energy used in the building to primary energy and CO_2 emissions.

8. Develop a common procedure for an energy performance certificate.

9. Develop and compile relevant standards applicable for each individual building category.

An example of the energy efficiency indication in buildings is highlighted in ISO 23045 as shown in Figure 7.5 [9].

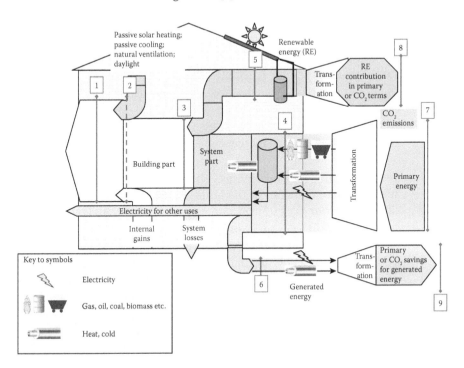

FIGURE 7.5
Global energy scheme (ISO 23045_2009). (From ISO publications ISO 23045_2009, January 2009.)

7.2.5 Concluding Remarks

It is important to incorporate an energy performance directive as a standard in this region; such a goal will aid energy savings in large buildings and will help set regulations for energy-efficient designs that are based on standard calculation methods. Recommendations include

1. Develop standardised tools for the calculation of the energy performance of buildings.

2. Define system boundaries for the different building categories and different heating systems.

3. Prepare models for expressing requirements on indoor air quality, thermal comfort in winter and when appropriate in summer, visual comfort and so forth.

4. Define comparable energy related key values (kWh/m², kWh per person, kWh per apartment, kWh per produced unit etc.) in order to develop a common procedure for an energy performance certificate.

5. Design, construct and operate a solar decathlon (building) that can meet the rural and desert requirements and save the diminishing fossil fuel sources

   a. Is the energy needed to fulfil the user's requirements for heating, lighting, cooling and so on, according to levels that are specified for the purposes of the calculation?

   b. Do the natural energy gains (passive solar, ventilation cooling, daylighting and so on) together with internal gains (occupants, lighting, electrical equipment, etc.) reduce energy demand in the winter season but increase energy demand in the summer season?

   c. Is the building's net energy use, obtained from (1) and (2) along with the characteristics of the building itself? [In winter season, (2) is lower than (1), but in summer, (2) is greater than (1).]

   d. Is the delivered energy, represented separately for each energy carrier, inclusive of auxiliary energy, used by heating, cooling, ventilation, hot water and lighting systems, taking into account renewable energy sources and cogeneration? This may be expressed in energy units or in units of the energy ware (kg, m³, kWh etc.).

   e. Is renewable energy produced on the building premises?

   f. Is generated energy produced on the premises and exported to the market? This can include part of (5).

   g. Does it represent the primary energy usage or the CO_2 emissions associated with the building.

h. Does it represent the primary energy or emissions associated with on-site generation that is used on-site and so is not subtracted from (7).

i. Does it represent the primary energy or CO_2 saving associated with exported energy, which is subtracted from (7).

7.3 New Design Practices

7.3.1 General

One of the Egyptian Department of Energy's greatest efficiency accomplishments is the Residential Air Conditioner Standards and Labelling programme. After years of coordination with manufacturers and the Department of Trade and Industry's EOS (Egyptian Standards Organisation), energy labelling was devised for air conditioners (whether split or window units, see Figure 7.6). Work is underway to incorporate energy labelling for the air-conditioning business. The programme has the potential to become a powerful platform for subsequent energy efficiency efforts not only in Egypt but also in other Arab countries. Air conditioners, both imported and domestic models, are required to meet a minimum efficiency standard and to be labelled. Although only in use in a small fraction of households, air conditioners are given priority because they represent one of the fastest growing electricity end uses in the residential sector. The impact of the programme will increase with time because the number of air conditioners in the country is rising dramatically.

Standards and labelling programmes provide enormous energy-saving potential that can direct developing countries toward sustainable energy use. Improved end-use efficiency programmes can contribute significantly to developing economies. The main benefits are

1. Less need to build new power plants
2. Reduced greenhouse gas emissions
3. Improved competitiveness for local manufacturers

FIGURE 7.6
Egyptian energy air conditioner label.

4. Higher consumer disposable income

5. Increased cash flow in the local economy

6. Improved trade balance by increased export earnings (or less import spending) helps alleviate trade deficit of Arabian countries

7. Avoid future energy deficit as power demand rises

7.3.2 Comfort Levels

7.3.2.1 Introduction

Thermal comfort is satisfaction with the thermal environment. Because there are large variations, both physiologically and psychologically from person to person, it is difficult to satisfy everyone in a space. The environmental conditions required for comfort are not the same for everyone. Extensive laboratory and field data have been collected that provide the necessary statistical data to define conditions that a specified percentage of occupants will find thermally comfortable. This is used to determine the thermal environmental conditions in a space that are necessary to achieve acceptance by a specified percentage of occupants of that space. There are six primary factors that must be addressed when defining conditions for thermal comfort. A number of other, secondary factors affect comfort in some circumstances [2].

The temperature regulatory centre in the brain is about 36.8°C at rest in comfort and increases to about 37.4°C when walking and 37.9°C when jogging [2]. High temperatures may cause increased out gassing of toxins from furnishings, finishes, building materials and so forth. Alternatively, ambient temperatures that are too cool can cause occupant discomfort such as shivering, inattentiveness and muscular and joint tension. Relative humidity plays an important role in the comfort feeling, affecting the comfort feeling directly or indirectly by its influence on the temperature. Excessive relative humidity levels are known to reduce human comfort. Most guidelines specify the range of 35%–50% as the optimum conditions for relative humidity. The airflow pattern plays an important role in the comfort sensation and also in the scavenging of the hazards and airborne particles. According to the results of research and the standards specifications, the optimum airflow velocity falls within the range of 0.2–0.25 m/s in the occupied zone.

7.3.2.2 Problem Identification

Many of the HVAC applications suffer from poor distribution of the indoor air temperature and relative humidity as well as from incorrect airflow velocities. This poor distribution arises from poor airflow distribution and the presence of thermal drift due to the buoyancy effect.

7.3.2.3 Status Quo

At present, most research recommends experimental and numerical simulation as the perfect tools to obtain the optimum design. The optimisation procedure of the HVAC airside design depends on the predictions of the air temperature distribution based on the simulation of different parametric designs using experimentally verified numerical tools. The influence of various ventilation strategies and vapour generation rate on the characteristics of temperature and moisture distribution is investigated [12,13]. The vast majority of the available thermal comfort data pertains to sedentary or near-sedentary physical activity levels typical of office work. The body of available data does not contain significant information regarding the comfort requirements of children, the disabled or the infirm.

Airside design and room furnishing were found influential in order to ensure a comfortable environment, especially in the displacement ventilation configuration [14]. The airflow velocity influence was investigated, and it was found that acceptable velocity could be as high as 0.35 m/s in the occupied zone [15]. The configurations of the conditioned space affect the comfort level as well as the space applications, such as healthcare facilities, which are so complex [16,17]. The correct specifications of outdoor ambient conditions affect thermal loads, as shown in Figure 7.1, and consequently, the comfort level [18]. Some research recommends changing the focus on the effect of building envelope to reduce the thermal load to enhance the thermal comfort [1,19–21].

7.3.2.4 Closure

The comfort conditions depend on many factors beyond the indoor air temperature, relative humidity and airflow velocity. Comfort conditions depend also on the air distribution pattern and the air movement [20], but the effect of these factors can be considered close to the air quality more than the comfort level. Indeed, the comfort criteria affect the air quality and the energy conservation in the ventilated and conditioned spaces, as shown in the following sections. The relation between the comfort and air quality is an interchange or a mutual relation.

7.3.3 Air Quality

7.3.3.1 Introduction

Most guidelines consider that the air quality is the result of a collaborative effort among environmental conditions (those presented in the introductory section). Indeed, in the present literature, the air quality is specified by a collaborative effort of the pressure relationship, air movement efficiency and contaminant concentration. These conditions play an important role in achieving optimum air quality. The ventilation system design must, as much as possible, provide air movement from the clean to the less clean areas. This rule requires great care in designing the airside system and in selecting the design of the airside system of neighbourhoods. There are relative interactions among the conditioned neighbouring spaces. The air distribution and movement efficiency can be considered as the simultaneous indicators of comfort and air quality. There are several important considerations that characterise the air distribution in air-conditioned spaces. First, the flow is generally turbulent, and buoyancy effects are often significant. Then the transverse transport effects are of particular interest in these flows. In that case, combined heat and mass transfer processes prevail, and coupled transport mechanisms are generally present. Undesirable airflow between rooms and floors is often difficult to control because of open doors, movement of staff and patients, temperature differentials and stack effect. Although some of these factors are beyond practical control, the effect of other factors may be minimised by terminating shaft openings in enclosed rooms and by designing and balancing air systems to create positive or negative air pressure within certain rooms and certain areas (Figure 7.7).

Contaminants can be classified in four broad headings, each of which represents a wide variety of pollutants: organic compounds, inorganic compounds, particulate matter and biological contaminants. It should be understood that these classifications are intended to facilitate the categorisation of contaminants. Although the pollutants are classified into these categories, certain contaminants may belong to two or more classifications,

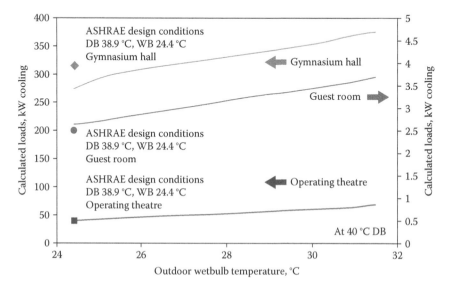

FIGURE 7.7
Outdoor conditions and thermal loads.

depending upon their nature. The classification of organic compounds represents chemical compounds that contain carbon–hydrogen bonds in their basic molecular structure. Their sources can be either natural products or synthetics, especially those derived from oil, gas and coal. Organic contaminants may exist in the form of gas (vapour), liquid or as solid particles in the atmosphere, food and/or water. Inorganic compounds are those that do not contain carbon–hydrogen bonds in their molecular structure.

The danger of particulate matter is its ability to become contaminated by other ambient sources, increasing health risks to individuals who are exposed to respirable suspended particles. Particles in this category are, usually, less than 10 mm in aerodynamic diameter. As mentioned previously, particles smaller than 5 mm are capable of bypassing the respiratory defences. Biological contaminants are generally referred to as microbes or microorganisms. Biological contaminants are minute particles of living matter produced from a variety of sources. The variety of biological compounds that may be present in the ambient environment is immense. Sources of pollution exist in the internal and external environment. The air quality is controlled by removal of the contaminant or by dilution. American Society of Heating, Refrigerating and Air conditioning Engineers (ASHRAE) [22] prescribes necessary quantities of ventilation for various types of occupancies and methods of determining the proportions of outside air and recirculated air. If the level of contaminants in outdoor air exceeds that for minimum air quality standards, extraordinary measures must be used. Although proper air-conditioning designs are helpful in the prevention and treatment of diseases, the application of air-conditioning to health facilities presents many specific problems. Those are not encountered in the conventional comfort conditioning design.

7.3.3.2 Problem Identification

The contaminant concentration mainly depends on two factors: air pressure relationship and air movement efficiency. So the optimum design of these two factors leads to accepted concentration and safe distribution of the contaminant. Actually, most guidelines to date do not restrict any airside design for each application. This gives a large tolerance and many design alternatives that are not totally perfect.

7.3.3.3 Status Quo

Comfort and air quality are investigated with the aid of experimental and numerical techniques to represent the relation between the thermal conditions and the air quality [1]. Thermal conditions affect the air quality; therefore, any recommended numerical models should account for balanced thermal conditions. This would affect the discrepancies between measured and simulated results and consequently create a more generalised numerical formula for the air characteristics. The effect of thermal loads and cooling strategies on the airflow pattern in an office was investigated by applying mixing ventilation [21] as well as those results indicating the effect of supply conditions on the airflow pattern [16]. The ventilation performance represents the capabilities of the airside design for providing a clean space. The effect of the heat and contaminant source location on the ventilation performance is also important. Optimum ventilation performance is achieved when these sources are located near the exhaust opening [22]. In healthcare applications, the air quality is generally influenced by airborne and contaminant generation, especially in critical sites such as the isolation and surgery rooms. Several studies have investigated airborne particle control in operating rooms using numerical techniques [23–26].

7.3.3.4 Closure

Air movement efficiency is mainly based on two factors: the air pressure relationship with the other neighbourhood spaces and the airside design. Differential air pressure can be maintained only in an entirely sealed room. Therefore, it is important to obtain a reasonably close fit of all doors and seal all walls and floor penetrations within pressurised areas. This is best accomplished by using weather stripping and drop bottoms on doors. The opening of a door between two areas instantaneously reduces any existing pressure differential between them to such a degree that its effectiveness is nullified. When such openings occur, a natural interchange of air takes place between the two rooms due to turbulence created by the door opening and closing combined with personal ingress/egress.

7.3.4 Energy-Efficient Building Design

7.3.4.1 Introduction

In the early 1970s, the energy crisis forced the development of conservation strategies in a variety of industries. Sustainability and energy efficiency continue to be strong issues in this time of limited resources. Therefore, the implementation of energy-conserving strategies in HVAC systems must be balanced with occupant comfort and health. A few guidelines gave specific recommendations about energy saving in HVAC systems, but these recommendations do not meet all requirements and design varieties. Indeed, in hot and humid climates, outdoor conditions play an important role in the energy consumption.

7.3.4.2 Problem Identification: Pyramid Concept

Until now, the guidelines and design standards have not provided restricted utilisation strategies for the conditioned air in spaces. This results in inefficient systems and expensive energy invoices. In some critical facilities, such as hospitals, HVAC designers face the problem of balancing healthy

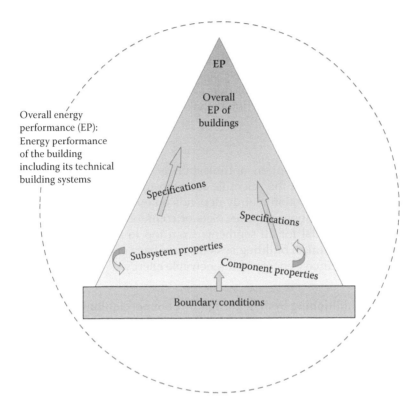

FIGURE 7.8
Overall building energy performance indicators.

conditions with energy utilisation. The assessment of the overall energy performance of a particular building, including the technical building systems, comprises a number of successive steps, which can be schematically visualised as a pyramid. Figure 7.8 shows various energy performance indicators of buildings.

7.3.4.3 Status Quo

The relation between the HVAC system design and optimum conditions and energy utilisation is still under investigation even today. Recent research [27] has investigated the effect of ventilation design on comfort and energy. The effect of displacement ventilation on the humidity gradient in a factory located in a hot and humid region has also been illustrated [27]. It was found that there is a strong dependence relationship between the correct supplying conditions and comfort. Indeed, displacement ventilation is recommended as an energy-efficient system, because this system gives the designers the suitable tolerance to select more economical supply conditions. In recent years, new ventilation system designs, such as under-floor ventilation systems, are slowly overcoming the problems of current systems. Under-floor air supply is recommended as an alternative to ceiling air supply in office buildings to overcome the lack of flexibility in ceiling systems and to improve comfort conditions [27,28]. The under-floor system is recommended it due to its capability of reducing energy consumption due to the operational characteristics of supplied air.

7.3.4.4 Closure

Because energy consumption optimisation is the new trend, new investigations are needed in the scientific research in order to achieve success. Actually, energy utilisation mainly depends on optimum utilisation of conditioned air in conditioned spaces. Sets of common terms, definitions and symbols are essential for all segments from top to bottom. These include energy needs, technical building systems, auxiliary energy use, recoverable system losses, primary energy and renewable energy.

7.3.5 Air-Conditioning System Design of Commercial Buildings

In theory, if properly applied, every system can be successful in any building. However, in practice, such factors as initial and operating costs, space allocation, architectural design, location and the engineer's evaluation and experience limit the proper choices for a given building type. Heating and air-conditioning systems should be simple in design and of proper size for a given building, should involve generally fairly low maintenance and low operating costs and should have optimal inherent thermal control as is economically possible. Such control might include materials with high thermal properties, insulation and multiple or special glazing and shading devices.

An example of commercial building application is provided here for libraries and museums. In general, libraries have storage areas, working and office areas, a main circulation desk, reading rooms, rare book vaults and small study rooms. In general, museums have exhibit areas, work areas, back offices and storage areas. Some larger museums may have souvenir shops, a restaurant or cafeterias and so forth.

7.3.5.1 Load Characteristics

Many libraries, especially college libraries, operate up to 12 hours per day and may run the air-conditioning equipment about 4200 hours per year. Such constant usage requires the selection of heavy duty, long-life equipment that requires little maintenance. Museums are generally open about 8–10 hours per day, 5–7 days per week. The ambient conditions should not vary in temperature or relative humidity. The conditions should remain constant 24 hours per day, year-round. Cold or hot walls and windows and hot steam or water pipes should be avoided. Object humidity may be destructive, even if the ambient relative humidity is under control.

1. Sun gain: Libraries and museums usually have windows, sometimes of stained glass and skylights—more are in traffic areas than in book stacks or storage areas. Care must be taken to minimise the effects of the sun; shortwave (actinic) rays are particularly injurious. Heat gain from skylights, often over artificially lighted frosted glass ceilings, can be reduced by a separate forced ventilation system.

2. Transmission: In winter, effects on objects located close to outside walls and possible condensation of moisture on the objects and the surface of outside walls must be evaluated. In summer, possible radiant effects from exposure should be considered.

3. People: Some areas may have concentrations as high as 1.0 m² per person, although office space will have closer to 10 or 15 m² per person, and book stack areas up to 100 m² per person.

4. Lights: Careful analyses of the required lighting intensity should be made in various rooms and in view of daylighting availability.

5. Stratification: In reading rooms, large entrance halls and large art galleries with high natural or false ceilings, air temperature may stratify.

7.3.5.2 Design Concepts

All air ducted systems are preferred in library public areas, but careful evaluation of relative humidity is essential. This is also true for museums because exhibit items are generally irreplaceable. In museums, people loads vary, depending on whether there is a new exhibit and on the time of day, the weather and other factors. Thus, individually controlled zones are required to maintain optimal environmental conditions. Attempts to establish a modular system for partitions have been only partially successful because of the wide range of sizes of items in the exhibits. In art museums, particularly, partitions may create local pockets with hot air supply or exhaust; transfer grilles may be placed in the partitions to obtain some air flow movement. Another problem is the location of room thermostats and humidistats.

7.3.5.3 Special Considerations

Many old manuscripts, books and artefacts [11,26–31] have been damaged or destroyed because they were not kept in a properly air-conditioned environment. The need for better preservation of such valuable materials, together with a rising popular interest in the use of libraries and museums, requires that most of them, whether new or existing, be air-conditioned. Air-conditioning problems for museums and libraries are generally similar but differ in design concept and application. Figure 7.9 depicts an example of artefact deterioration due to excessive humidity [26,32].

FIGURE 7.9
Effect of moisture content on the colour of artefacts.

7.3.5.4 Design Criteria

In an average library or museum, less stringent design criteria are usually provided than for archives because the value of the books and collections does not justify the higher initial and operating costs. Low-efficiency air filters are often provided. Relative humidity is held below 55%. Room temperatures are held within the 20–21.5°C range. Archival libraries and museums should have 85% or better air filtration, a relative humidity of 35% for books and temperatures of 16°C in book stacks and 20°C in reading rooms.

7.3.5.5 Building Contents

The reaction of museum contents and collections to room conditions should be carefully considered and critically examined. For example, paper used in books and manuscripts prior to the 18th century is very stable and is not significantly affected by the room environment. For archival preservation, this paper should be stored at very low temperatures. It is estimated that for each 5°C dry bulb the room temperature is lowered, the life of the paper will double and that any humidity reduction will also lengthen the life of paper.

7.3.5.6 Effect of Ambient Atmosphere

The temperature and, particularly, the relative humidity of the air have a marked influence on the appearance, behaviour and general quality of hygroscopic materials such as paper, textiles, wood and leather because the moisture content of these substances comes into equilibrium with the moisture content of the surrounding air. The object humidity is usually defined as the relative humidity of the thin film of air in close contact with the surface of an object and at a temperature cooler or warmer than the ambient dry bulb. If artefacts are permitted to cool overnight, the next day they will be enveloped by layers of air having progressively higher relative humidity. This may range from the ambient of 45% to 60% to 97% immediately next to the object surface, thus effecting a change in material regain or even condensation. This, combined with the hygroscopic or salty dust often found on objects recovered from excavations, can be destructive.

7.3.5.7 Sound and Vibration

Air-conditioning equipment should be treated with sound and vibration isolation to ensure quiet comfort for visitors and staff as per the ASHRAE standards and local environmental laws.

7.3.6 Evaluation Indices

7.3.6.1 Introduction

The evaluation indices of the comfort, air quality and energy utilisation efficiency can be divided into two main categories—empirical indices based on experimental techniques and numerical indices based on numerical techniques. The most common indices provide the required evaluation of the air characteristics at individual positions (or in other scope, at individual points) in the indoor environment.

7.3.6.2 Problem Identification

Until now, the evaluation of the comfort, air quality and energy utilisation efficiency was performed only at individual positions (local evaluation). There still is no general global evaluation index for several characteristics such as the airflow movement and the contaminant concentration and its influence on the occupancy health. Actually, the air flow distribution pattern plays the role of the global evaluation index up today. On the other hand, there is no global evaluation index capable of evaluating comfort, air quality and energy utilisation efficiency simultaneously. Actually, this global index will aid the HVAC designers in achieving the optimum design according to the optimum indoor air quality levels [27–29].

7.3.6.3 Status Quo

Energy efficiency is better characterised in buildings through the energy efficiency ratio (EER), an index that is mandatory in all air-conditioning and refrigeration systems. The EER is defined as the useful output divided by the energy input, a form of coefficient of performance. International standards dictate that EER is greater than a pre-set value in an energy label (as shown here in Figure 7.10) [33]. It is classified in efficiency categories from A to G, giving the cooling output at full load in kW and the EER in cooling mode at full load.

Countries set their own energy labels based on local industry and on energy generation potentials and strategies. The minimum EER in Egypt, for example, is 2.65 for split air conditioners [33]. The minimum efficiencies

Air conditioners, cooling EER in W/W

A	B	C	D	E	F	G
>3.2	3.0–3.2	2.8–3.0	2.6–2.8	2.4–2.6	2.2–2.4	<2.2

(a)

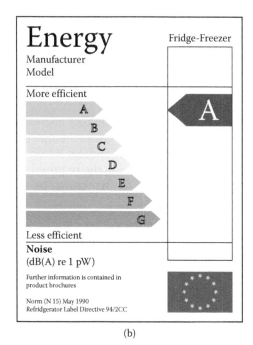

(b)

FIGURE 7.10
Examples of energy efficiency ratio for products. (a) Energy efficiency ratio classifications for room air conditioners. (b) Energy efficiency ratio classifications for refrigerators. (From European Commission Energy Labeling, 2010.)

of mechanical systems are commonly known as MEPS and are also set to rationalise the energy use in fossil fuel equipment and to reduce the carbon footprint. Assessment of energy performance of buildings as a whole are generally use energy rating systems; one example is the Green Pyramids Rating System (GRPS) [33], Leadership in Energy and Environmental Design (LEED) [34].

The GPRS Green Pyramid Category Weightings as specified are as follows:

1. Sustainable site, accessibility and ecology   10%
2. Energy efficiency   20%
3. Water utilisation efficiency   30%
4. Materials and resources   10%

5. Indoor environmental quality	10%
6. Management	10%
7. Innovation and added value	10%

The first group of assessment points goes to site and ecology. These were set to encourage development in desert areas and redevelopment in informal areas and to avoid projects that negatively affect archaeological, historical and protected areas. This is also to minimise pollution and traffic congestion from car use and to conserve non-renewable energy by encouraging public and alternative transport. Ultimately, this would minimise the environmental impact of the project on the site and its surroundings; it would protect existing natural systems, such as fauna and flora (including wildlife corridors and seasonal uses), soil, hydrology and groundwater from damage and would promote biodiversity. Ten points were given for that group, while 20 points were given for energy-efficient building designs utilising the concepts highlighted here and by Van Dijk and Khalil [11,31] and Ashrae [29]. The water energy nexus was realised in a 30 point assessment for the water efficiency procedures. Materials and resources were given 10 points as well. Indoor environmental quality as dedicated by human thermal comfort and acoustics and by natural and artificial lighting was given 10 points as well. Proper management and innovation harvest the remaining 20 points. A balance of all the important and influential factors was accounted for with these rating systems. The building would be certified green if attaining a minimum of 80 points and would just be certified if attaining up to 49 points, as indicated in Figure 7.11.

The final credits are calculated and are categorised within the following rating:

1. GPRS Certified: 40–49 credits
2. Silver Pyramid: 50–59 credits
3. Gold Pyramid: 60–79 credits
4. Green Pyramid: 80 credits and above

7.3.6.4 Closure

Because optimisation of energy consumption is the new trend, the achievement of this level needs new investigation in scientific studies. Actually, energy utilisation mainly depends on the optimum utilisation of conditioned air in conditioned spaces. Sets of common terms, definitions and symbols are essential for all segments from top to bottom. The target of this section is to highlight procedures to control the alteration, repair, maintenance and operation of existing building sites and the alteration of building site improvements where additions are made to, or changes of occupancy occur within, the existing buildings on the site. Building sites shall be operated and maintained in conformance with the national green building code.

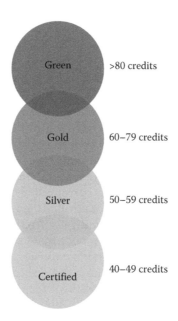

FIGURE 7.11
Green building rating system.

The owner or the owner's designated agent shall be responsible for the operation and maintenance of building sites. The requirements of the green building code shall be incorporated and implemented in new and renovation work. Alterations and repairs to building sites shall comply with the code provisions. Building materials used for building site development shall comply with the requirements of the code materials; systems already in use on a building site that are in conformance with the requirements or approvals in effect at the time of their installation shall be permitted to remain in use unless determined by the code official to be dangerous to the environment, life, health or safety. Where such conditions are determined to be dangerous to the environment, life, health or safety, they shall be mitigated or made safe.

References

1. Khalil, E. E. (2005). Energy performance of buildings directive in Egypt: A new direction, *HBRC Journal*, Vol. 1, pp. 197–213.
2. Medhat, A. A. and Khalil, E. E. (2006). Thermal comfort meets human acclimatization in Egypt, *Proceeding of Healthy Building*, Vol. 2, p. 25.
3. Khalil, E. E. (2006). Energy performance of commercial buildings in Egypt: A new direction, *Proceedings, Energy2030*, Abu Dhabi, November 2006.

4. Kosonen, R. (2002). Displacement ventilation for room air moisture control in hot and humid climate, *ROOMVENT 2002*, pp. 241–244.

5. Leite, B. C. C. and Tribess, A. (2002). Analysis of under floor air distribution system: Thermal comfort and energy consumption, *ROOMVENT 2002*, pp. 245–248.

6. Kameel, R. and Khalil, E. E. (2002). Prediction of turbulence behavior using k-ε model in operating theatres, *ROOMVENT 2002*, pp. 73–76.

7. Khalil, E. E. (2008). Arab-air conditioning and refrigeration code for energy-efficient buildings, *Arab Construction World*, Vol. 28, No. 8, pp. 24–26.

8. Khalil, E. E. (2008). Air conditioning and refrigeration code for energy-efficient buildings in the Arab world, *Journal of Kuwait Society of Engineers*, Vol. 100, pp. 94–95.

9. ISO publications ISO 23045_2009, January 2009.

10. Federal Register/Vol. 72, No. 245/Friday, December 21, 2007/Rules and Regulations 72565.

11. Van Dijk, D. and Khalil, E. E. (2011). Future cities—Building on energy efficiency, *ISO Focus*, pp. 25–27.

12. EPD, European Energy Performance Code, CEN, 2006.

13. Olesen, B. W., Koschenz, M. and Johansson, C. (2003). New European standard proposal for design and dimensioning of embedded radiant surface heating and cooling systems, *ASHRAE Transactions*, Vol. 109, p. KC-03-07-4.

14. Kameel, R. and Khalil, E. E. (2003). Energy efficiency, merits, and advantages of various air-conditioning system in commercial buildings, in Egypt, *Proceedings of ASHRAE-RAL, Paper Ral.3-5*, Cairo.

15. Khalil, E. E. (2003). HVAC in energy efficiency building code, Egypt, *Proceedings of ASHRAE-RAL, Paper Ral.3-6*, Cairo.

16. PrEN ISO 10077-2, Thermal performance of windows, doors and shutters—Calculation of thermal transmittance—Part 2: Numerical methods for frames, 2005.

17. PrEN 13363-1, Solar protection devices combined with glazing—Calculation of solar energy and light transmittance—Part 1: Simplified method, 2005.

18. ISO 13790, Thermal performance of buildings—Calculation of energy use for space heating, 2010.

19. Khalil, E. E. (2004). Energy efficiency in air conditioned buildings: An overview, *Proceedings of 6th JIMEC*, Amman.

20. Khalil, E. E. (2004). Energy efficiency in air conditioned buildings in Egypt, *Proceedings of the 6th JIMEC*, Amman.

21. USA DOE (2002). Department of Energy.

22. *ASHRAE Handbook, Fundamentals 2013*, ASHRAE, Atlanta, GA.

23. Naydenov, K., Pitchurov, G., Langkilde, G. and Melikov, A. K. (2002). Performance of displacement ventilation in practice, *ROOMVENT 2002*, pp. 483–486.

24. Jacobsen, T. S., Hansen, R., Mathiesen, E., Nielsen, P. V. and Topp, C. (2002). Design method and evaluation of thermal comfort for mixing and displacement ventilation, *ROOMVENT 2002*, pp. 209–212.

25. Kameel, R. and Khalil, E. E. (2001). Numerical computations of the fluid flow and heat transfer in air-conditioned spaces, *NHTC2001-20084, 35th National Heat Transfer Conference*, Anaheim, CA.

26. Khalil, E. E. (2009). Thermal management in hospitals: Comfort, air quality and energy utilization, *Proceedings ASHRAE, RAL*, Kuwait.

27. Nakamura, Y. and Fujikawa, A. (2002). Evaluation of thermal comfort and energy conservation of an ecological village office, *ROOMVENT 2002*, pp. 413–416.
28. ASHRAE, Applications, 2011, ASHRAE, Atlanta, GA.
29. ASHRAE standards 55-2013, ASHRAE, Atlanta, GA.
30. Holmberg, S. and Einberg, G. (2002). Flow behavior in a ventilated room—Measurements and simulations, *ROOMVENT 2002*, pp. 197–200.
31. Van Dijk, D. and Khalil, E. E. (2009). Energy efficiency in buildings, *ISO Focus*, September 2009, pp. 16–20.
32. Khalil, E. E. (2011). Ventilation of the archaeological tombs of the valley of kings, Luxor, Egypt, *Proceedings of International Conference on Air-Conditioning & Refrigeration 2011*, ICACR2011, ICACR2011-00102, Korea.
33. European Commission Energy Labeling, 2010, at www.epbd.eu.
34. Khalil, E. E. (2012). International focus on emerging technologies & opportunities, *ASME Congress*, Paper Number IMECE2012-94127.

Further Readings

1. ASHRAE Guideline 1.1—HVAC&R Technical Requirements for The Commissioning Process, Latest Edition.
2. ASHRAE Standard 52.2—Method of Testing General Ventilation Air-Cleaning Devices for Removal Efficiency by Particle Size, Latest Edition 2007.
3. ASHRAE Standard 55—Thermal Environmental Conditions for Human Occupancy, Latest Edition.
4. ASHRAE Standard 62.1—Ventilation for Acceptable Indoor Air Quality—Sets the minimum acceptable ventilation requirements, Latest Edition.
5. ASHRAE Standard 90.1—Energy Standard for Buildings Except Low-Rise Residential Buildings.
6. Department of Defense.
7. Air Force Engineering Technical Letter ETL04-3 Design Criteria for Prevention of Mold in Air Force Facilities, by U.S. Air Force, 2004.
8. U.S. General Services Administration.
9. P100 Facilities Standards for the Public Buildings Service, 2010.

8

Green Buildings

8.1 General

Green construction (or sustainable building) is a structural process that is environmentally responsible and resource efficient throughout a building's life cycle, from laying down the design principles, construction, operation, maintenance, renovation, to demolition. This requires close cooperation among the design team, the architects, the engineers and the client at all project stages. Green building practices expand and complement the classical building design concerns of economy, utility, durability and comfort.

Although new technologies are constantly being developed to complement current practices in creating greener structures, the common objective is that green buildings are designed to reduce the overall impact of the built environment on human health and on the natural environment by

1. Efficiently using energy, water and other resources
2. Protecting occupant health and improving employee productivity
3. Reducing waste, pollution and environmental degradation

A similar concept is natural building, which is usually on a smaller scale and tends to focus on the use of natural materials that are available locally [1]. Other related topics include sustainable design and green architecture. Sustainability may be defined as meeting the needs of present generations without compromising the ability of future generations to meet their own needs. Although some green building programmes do not address the issue of retrofitting existing homes, others do. Green construction principles can easily be applied to retrofit work as well as to new construction.

A 2009 report by the U.S. General Services Administration [2–4] found 12 sustainably designed buildings cost less to operate and had excellent energy performance. In addition, occupants were more satisfied with the overall building than were those in typical commercial buildings.

Although the definition of *sustainable building design* is constantly changing, six fundamental principles persist.

1. Optimise site potential: Creating sustainable buildings starts with proper site selection, including consideration of the reuse or rehabilitation of existing buildings. The location, orientation and landscaping of a building affects local ecosystems, transportation methods and energy use. It is important to incorporate smart growth principles into the project development process, whether the project is a single building, a campus or a military base. Siting for physical security is a critical issue in optimising site design, including locations of access roads, parking, vehicle barriers and perimeter lighting. Whether designing a new building or retrofitting an existing building, site design must integrate with sustainable design to achieve a successful project. The site of a sustainable building should reduce, control and/or treat storm water runoff. If possible, planners should strive to support native flora and fauna of the region in the landscape design.

2. Optimise energy use: With the continually increasing demand on the world's fossil fuel resources, concerns for energy independence and security are increasing, and the impacts of global climate change are becoming more evident. Therefore, it is essential to find ways to reduce energy load, increase efficiency and maximise the use of renewable energy sources in federal facilities. Improving the energy performance of existing buildings is important to increasing our energy independence. Government and private sector organisations are increasingly committed to building and operating net zero energy buildings as a way to significantly reduce our dependence on fossil fuel-derived energy.

3. Protect and conserve water: In many parts of the country, fresh water is an increasingly scarce resource. A sustainable building should use water efficiently and should reuse or recycle water for on-site use, when feasible. The effort to bring drinkable water to our household faucets consumes enormous energy resources in pumping, transport and treatment. Often potentially toxic chemicals are used to make water potable. The environmental and financial costs of sewage treatment are significant.

4. Optimise building space and material use: The materials used in a sustainable building minimise life cycle environmental impacts such as global warming, resource depletion and human toxicity. Environmentally preferable materials have a reduced effect on human health and the environment and contribute to improved worker safety and health, reduced liabilities, reduced disposal costs and achievement of environmental goals.

5. Enhance indoor environmental quality: The indoor environmental quality (IEQ) [2–4] of a building has a significant impact on occupant

health, comfort and productivity. Among other attributes, a sustainable building maximises day lighting, has appropriate ventilation and moisture control, optimises acoustic performance and avoids the use of materials with high-volatile organic compounds (VOC) emissions. Principles of IEQ also emphasise occupant control over systems such as lighting and temperature.

6. Optimise operational and maintenance practices: Considering a building's operating and maintenance issues during the preliminary design phase of a facility will contribute to improved working environments, higher productivity and reduced energy and resource costs, and will prevent system failures. Encourage building operators and maintenance personnel to participate in the design and development phases to ensure optimal operations and maintenance of the building. Designers can specify materials and systems that simplify and reduce maintenance requirements; that require less water, energy, and toxic chemicals and cleaners to maintain; and that are cost-effective and reduce life cycle costs. Additionally, design facilities to include meters in order to track the progress of sustainability initiatives, including reductions in energy and water use and waste generation, in the facility and on site.

8.2 Concepts

The concept of sustainable development can be traced to the energy (especially fossil oil) crisis and the environment pollution concerns in the 1970s. The U.S. green building movement originated with the need and desire for more energy-efficient and environment-friendly construction practices. There are a number of motives for building green, including environmental, economic and social benefits. However, modern sustainability initiatives call for an integrated and synergistic design in new construction and in the retrofitting of existing structures. Also known as sustainable design, this approach integrates the building life cycle with each green practice employed following a design-purpose of creating a synergy among the practices used.

Green building brings together a vast array of practices, techniques and skills to reduce and ultimately eliminate the impacts of buildings on the environment and on human health. It often emphasises taking advantage of renewable resources, such as using sunlight through passive solar, active solar, and photovoltaic equipment, and using plants and trees through green roofs, rain gardens and reduction of rainwater runoff. Many other techniques are used, such as using low-impact building materials or using packed gravel or permeable concrete instead of conventional concrete or asphalt to enhance replenishment of ground water.

Although the practices or technologies employed in green building are constantly evolving and may differ from region to region, fundamental principles persist from which the method is derived: sitting and structure design efficiency, energy efficiency, water efficiency, materials efficiency, IEQ enhancement, operations and maintenance optimisation, and waste and toxics reduction. The essence of green building is an optimisation of one or more of these principles. Also, with the proper synergistic design, individual green building technologies may work together to produce a greater cumulative effect.

On the aesthetic side of green architecture or sustainable design is the philosophy of designing a building that is in harmony with the natural features and resources surrounding the site. There are several key steps in designing sustainable buildings: specify 'green' building materials from local sources, reduce loads, optimise systems and generate on-site renewable energy.

8.2.1 Life Cycle Assessment

A life cycle assessment (LCA) can help avoid a narrow outlook on environmental, social and economic concerns by assessing a full range of impacts associated with all cradle-to-grave stages of a process: from extraction of raw materials through materials processing, manufacture, distribution, use, repair and maintenance, and disposal or recycling [5,6]. Impacts taken into account include (among others) embodied energy, global warming potential, resource use, air pollution, water pollution and waste.

In terms of green building, the last few years have seen a shift away from a *prescriptive* approach, which assumes that certain prescribed practices are better for the environment, toward the scientific evaluation of actual performance through LCA.

Although LCA is widely recognised as the best way to evaluate the environmental impacts of buildings (ISO 14040 provides a recognised LCA methodology), it is not yet a consistent requirement of green building rating systems and codes, despite the fact that embodied energy and other life cycle impacts are critical to the design of environmentally responsible buildings.

In North America, LCA is rewarded to some extent in the Green Globes® rating system, and it is part of the new American National Standard based on Green Globes, *ANSI/GBI 01-2010: Green Building Protocol for Commercial Buildings*. LCA is also included as a pilot credit in the Leadership in Energy and Environmental Design (LEED) system, though a decision has not been made as to whether it will be incorporated fully into the next major revision. The state of California also included LCA as a voluntary measure in its 2010 draft *Green Building Standards Code*.

Although LCA is often perceived as overly complex and time consuming for regular use by design professionals, research organisations such as BRE in the UK and the Athena Sustainable Materials Institute in North America are working to make it more accessible.

In the UK, the BRE *Green Guide to Specifications* offers ratings for 1,500 building materials based on LCA.

In North America, the ATHENA® *EcoCalculator for Assemblies* provides LCA results for several hundred common building assembles based on data generated by its more complex parent software, the ATHENA® *Impact Estimator for Buildings*. (The EcoCalculator is available free at www.athenasmi.org.) Athena software tools are especially useful early in the design process when material choices have far-reaching implications for overall environmental impact. They allow designers to experiment with different material mixes to achieve the most effective combination.

8.3 Measures and Assessments

Measures and assessments of green building performance and energy efficiency can be effected through

1. Design considerations and enforcements
2. Code compliance
3. Post-occupancy assessment

The five guiding principles with which buildings must comply are as follows:

1. Employ integrated assessment, operation and management principles
2. Optimise energy performance
3. Protect and conserve water
4. Enhance IEQ
5. Reduce environmental impact of materials

8.3.1 Guiding Principles for Sustainable Existing Buildings

1. Employ Integrated Assessment, Operation and Management Principles
    a. *Integrated assessment, operation and management.* Use an integrated team to develop and implement policy regarding sustainable operations and maintenance.
        i. Incorporate sustainable operations and maintenance practices within the appropriate Environmental Management System.

    ii. Assess existing condition and operational procedures of the building and major building systems and identify areas for improvement.

   iii. Establish operational performance goals for energy, water, material use and recycling, and IEQ, and ensure incorporation of these goals throughout the remaining life cycle of the building.

   iv. Incorporate a building management plan to ensure that operating decisions and tenant education are carried out with regard to integrated, sustainable building operations and maintenance.

   v. Augment building operations and maintenance as needed using occupant feedback on work space satisfaction.

  b. *Commissioning.* Employ recommissioning, tailored to the size and complexity of the building and its system components, in order to optimise and verify performance of fundamental building systems. Commissioning must be performed by an experienced commissioning provider. When building commissioning has been performed, the commissioning report, summary of actions taken and schedule for recommissioning must be documented. Building recommissioning must have been performed within four years prior to reporting a building as meeting the guiding principles.

2. Optimise Energy Performance

  a. *Energy efficiency.* Three options can be used to measure energy efficiency performance:

    i. Option 1: Receive an ENERGY STAR® rating of 75 or higher or an equivalent Labs21 Benchmarking Tool score for laboratory buildings.

   ii. Option 2: Reduce measured building energy use by 20% compared to building energy use in 2003 or a year thereafter with quality energy use data.

   iii. Option 3: Reduce energy use by 20% compared to the ASHRAE 90.1-2007 baseline building design if design information is available. Use ENERGY STAR and FEMP-designated energy-efficient products, where available.

  b. *On-site renewable energy.* Per E.O. 13423, implement renewable energy generation projects on agency property for agency use, when life cycle is cost-effective.

  c. *Measurement and verification.* Per the Energy Policy Act of 2005 Section 103, install building level electricity meters to track and continuously optimise performance. Per the EISA 2007, the utility meters must also include natural gas and steam, where natural gas and steam are used.

d. *Benchmarking.* Compare annual performance data with previous years' performance data, preferably by entering annual performance data into the ENERGY STAR Portfolio Manager. For building and space types not available in ENERGY STAR, use an equivalent benchmarking tool such as the Labs21 benchmarking tool for laboratory buildings.

3. Protect and Conserve Water

a. *Indoor water.* Two options can be used to measure indoor potable water use performance:

    i. Option 1: Reduce potable water use by 20% compared to a water baseline calculated for the building. The water baseline, for buildings with plumbing fixtures installed in 1994 or later, is 120% of the Uniform Plumbing Codes 2012 or the International Plumbing Codes 2012 fixture performance requirements. The water baseline for plumbing fixtures older than 1994 is 160% of the Uniform Plumbing Codes 2012 or the International Plumbing Codes 2006 fixture performance requirements.

    ii. Option 2: Reduce building measured potable water use by 20% compared to building water use in 2003 or a year thereafter with quality water data.

b. *Outdoor water.* Three options can be used to measure outdoor potable water use performance:

    i. Option 1: Reduce potable irrigation water use by 50% compared to conventional methods.

    ii. Option 2: Reduce building-related potable irrigation water use by 50% compared to measured irrigation water use in 2003 or a year thereafter with quality water data.

    iii. Option 3: Use no potable irrigation water.

c. *Measurement of water use.* The installation of water meters for building sites with significant indoor and outdoor water use is encouraged. If only one meter is installed, reduce potable water use (indoor and outdoor combined) by at least 20% compared to building water use in 2003 or a year thereafter with quality water data. Employ strategies that reduce storm water runoff and discharges of polluted water offsite. Per EISA Section 438, where redevelopment affects site hydrology, use site planning, design, construction and maintenance strategies to maintain hydrologic conditions during development, or to restore hydrologic conditions following development, to the maximum extent that is technically feasible.

d. *Process water.* Per the EP Act 2005 Section 109, when potable water is used to improve a building's energy efficiency, deploy life cycle cost-effective water conservation measures.

e. *Water-efficient products.* Where available, use EPA's WaterSense-labelled products or other water-conserving products, where available. Choose irrigation contractors who are certified through a WaterSense-labelled programme.

4. Enhance IEQ

a. *Ventilation and thermal comfort.* Meet ASHRAE Standard 55-2013, Thermal Environmental Conditions for Human Occupancy and ASHRAE Standard 62.1-2010: Ventilation for Acceptable Indoor Air Quality.

b. *Moisture control.* Provide policy and illustrate the use of an appropriate moisture control strategy to prevent building damage, minimise mould contamination and reduce health risks related to moisture. For facade renovations, Dew Point analysis and a plan for clean up or infiltration of moisture into building materials are required.

c. *Daylighting and lighting controls.* Automated lighting controls (occupancy/vacancy sensors with manual-off capability) are provided for appropriate spaces including restrooms, conference and meeting rooms, employee lunch and break rooms, training classrooms and offices. Two options can be used to meet additional daylighting and lighting controls performance expectations:

    i. Option 1: Achieve a minimum daylight factor of 2% (excluding all direct sunlight penetration) in 50% of all space occupied for critical visual tasks.

    ii. Option 2: Provide occupant controlled lighting, allowing adjustments to suit individual task needs, for 50% of regularly occupied spaces.

d. *Low-emitting materials.* Use low-emitting materials for building modifications, maintenance and cleaning. In particular, specify the following materials and products to have low pollutant emissions: composite wood products, adhesives, sealants, interior paints and finishes, solvents, carpet systems, janitorial supplies and furnishings.

e. *Integrated pest management.* Use integrated pest management techniques as appropriate to minimise pesticide usage. Use EPA-registered pesticides only when needed.

    f. *Environmental tobacco smoke control.* Prohibit smoking within the building and within 25 feet of all building entrances, operable windows and building ventilation intakes.

5. Reduce Environmental Impact of Materials

    a. *Recycled content.* Per Section 6002 of the Resource Conservation and Recovery Act (RCRA), for EPA-designated products, use products meeting or exceeding EPA's recycled content recommendations for building modifications, maintenance and cleaning. For other products, use materials with recycled content such that the sum of post-consumer recycled content plus one-half of the pre-consumer content constitutes at least 10% (based on cost or weight) of the total value of the materials in the project. If EPA-designated products meet performance requirements and are available at a reasonable cost, a preference for purchasing them shall be included in all solicitations relevant to construction, operation, maintenance of or use in the building. EPA's recycled content product designations and recycled content recommendations are available on relevant websites.

    b. *Biobased content.* Per Section 9002 of the Farm Security and Rural Investment Act (FSRIA), for USDA-designated products, use products with the highest content level per USDA's biobased content recommendations. For other products, use biobased products made from rapidly renewable resources and certified sustainable wood products. If these designated products meet performance requirements and are available at a reasonable cost, a preference for purchasing them should be included in all solicitations relevant to construction, operation, maintenance of or use in the building.

    c. *Environmentally preferable product.* Use products that have a lesser or reduced effect on human health and the environment over their life cycle when compared with competing products or services that serve the same purpose. A number of standards and eco labels are available in the marketplace to assist specifiers in making environmentally preferable decisions.

    d. *Waste and materials management.* Provide reuse and recycling services for building occupants, where markets or on-site recycling exist. Provide salvage, reuse and recycling services for waste generated from building operations, maintenance, repair and minor renovations, and discarded furnishings, equipment and property. This could include such things as beverage containers and paper from building occupants, batteries, toner cartridges, outdated computers from an equipment update and construction materials from a minor renovation.

e. *Ozone-depleting compounds.* Eliminate the use of ozone-depleting compounds where alternative environmentally preferable products are available, consistent with either the Montreal Protocol and Title VI of the Clean Air Act Amendments of 1990, or equivalent overall air quality benefits that take into account life cycle impacts.

8.3.2 Design Considerations and Enforcement

During the design stage of buildings projects that require and use energy extensively, the designers should follow all known techniques, options and codes to end up with an energy-efficient design.

Rating systems are groups of requirements for projects that want to achieve LEED certification. Each group is geared toward the unique needs of a project or building type.

LEED is flexible enough to apply to all project types including healthcare facilities, schools, homes and even entire neighbourhoods. The rating system selection guidance will help you determine which specific rating system is right for your project.

Projects earn points to satisfy green building requirements.

Within each of the LEED credit categories, projects must satisfy prerequisites and earn points. The number of points the project earns determines its level of LEED certification. The main credit categories are as follows.

1. *Sustainable site credits* encourage strategies that minimise the impact on ecosystems and water resources.

2. *Water efficiency credits* promote smarter use of water, inside and out, to reduce potable water consumption.

3. *Energy and atmosphere credits* promote better building energy performance through innovative strategies.

4. *Materials and resources credits* encourage using sustainable building materials and reducing waste.

5. *IEQ credits* promote better indoor air quality and access to daylight and views.

Additional LEED for Neighbourhood Development credit categories are

1. Smart location and linkage credits promote walkable neighbourhoods with efficient transportation options and open space.

2. Neighbourhood pattern and design credits emphasise compact, walkable, vibrant, mixed-use neighbourhoods with good connections to nearby communities.

3. Green infrastructure and buildings credits reduce the environmental consequences of the construction and operation of buildings and infrastructure.

Additional LEED for Homes credit categories are:

1. Location and linkage credits encourage construction on previously developed or infill sites and promote walkable neighbourhoods with access to efficient transportation options and open space.
2. Awareness and education credits encourage home builders and real estate professionals to provide homeowners, tenants and building managers with the education and tools they need to understand and make the most of the green building features of their home.

In addition, there are two bonus credit categories:

1. Innovation in design or innovation in operations credits addresses sustainable building expertise as well as design measures not covered under the five LEED credit categories. Six bonus points are available in this category.
2. Regional priority credits address regional environmental priorities for buildings in different geographic regions. Four bonus points are available in this category.

8.4 New Design Practices and Renewable Energy Blends

Various countries currently rely heavily on coal, oil and natural gas for their energy. Fossil fuels are non-renewable; that is, they draw on finite resources that will eventually dwindle, becoming too expensive or too environmentally damaging to retrieve. In contrast, the many types of renewable energy resources—such as wind and solar energy—are constantly replenished and will never run out. Most renewable energy comes either directly or indirectly from the sun. Sunlight, or solar energy, can be used directly for heating and lighting homes and other buildings, for generating electricity, and for hot water heating, solar cooling, and a variety of commercial and industrial uses (Figure 8.1).

Solar heat also drives the winds, whose energy is captured with wind turbines. Then the winds and the sun's heat cause water to evaporate. When this water vapour turns into rain or snow and flows downhill into rivers or streams, its energy can be captured using hydroelectric power.

Along with the rain and snow, sunlight causes plants to grow. The organic matter that makes up those plants is known as biomass. Biomass can be used to produce electricity, transportation fuels or chemicals. The use of biomass for any of these purposes is called bioenergy.

Hydrogen also can be found in many organic compounds as well as in water. It is the most abundant element on the Earth. But it does not occur

FIGURE 8.1
Solar shingles are installed on a rooftop.

naturally as a gas. It is always combined with other elements, such as with oxygen to make water. Once separated from another element, hydrogen can be burned as a fuel or can be converted into electricity.

Not all renewable energy resources come from the sun. Geothermal energy taps the Earth's internal heat for a variety of uses, including electric power production and the heating and cooling of buildings. And the energy of the ocean's tides come from the gravitational pull of the moon and the sun upon the Earth.

8.4.1 Ocean and Alternative Energies

In fact, ocean energy comes from a number of sources. In addition to tidal energy, there is the energy of the ocean's waves, which are driven by tides and winds. The sun also warms the surface of the ocean more than the ocean depths, creating a temperature difference that can be used as an energy source. All these forms of ocean energy can be used to produce electricity.

Why Is Renewable Energy Important?
Renewable energy technology content for this section provided in part by the National Renewable Energy Laboratory and the Department of Energy.

Cost-effective energy efficiency programmes have been delivered by large and small utilities and third-party programme administrators in some parts of the country since the late 1980s. The rationale for utility investment in efficiency programming is that within certain existing markets for energy-efficient products and services, there are barriers that can be overcome to ensure that customers from all sectors of the economy choose more energy-efficient products and practices. Successful programmes have developed strategies to overcome these barriers, in many cases partnering with industry and voluntary national and regional programmes so that efficiency programme spending is used not only to acquire demand-side resources but also to accelerate market-based purchases by consumers [4].

8.5 Closure

Energy efficient best practices can be summaries as

1. Recognise energy efficiency as a high-priority energy resource
2. Make a strong, long-term commitment to cost-effective energy efficiency as a resource
3. Broadly communicate the benefits of, and opportunities for, energy efficiency
4. Provide sufficient and stable programme funding to deliver energy efficiency where cost-effective

A list of options for promoting best practice energy efficiency appears at the end of this chapter. Challenges that limit greater utility investment in energy efficiency include the following:

1. The majority of utilities recover fixed operating costs and earn profits based on the volume of energy they sell.
2. There is a lack of standard approaches on how to quantify and incorporate the benefits of energy efficiency into resource planning efforts and institutional barriers at many utilities that stem from the historical business model of acquiring generation assets and building transmission and distribution systems. Strategies for overcoming these challenges were addressed by incorporating energy efficiency into resource planning.
3. Some rate designs that are counterproductive to energy efficiency might limit greater efficiency investment by large customer groups where effective opportunities for efficiency programming exist. Strategies for encouraging rate designs that are compatible with energy efficiency are needed.
4. Efficiency programmes need to address multiple customer needs and stakeholder perspectives while simultaneously addressing multiple system needs, in many cases while competing for internal resources.

This chapter focuses on strategies for making energy efficiency a resource, developing a cost-effective portfolio of energy efficiency programmes for all customer classes, designing and delivering efficiency programmes that optimise budgets, and ensuring that those programmes deliver results with the goal of creating a sustainable, aggressive national commitment to energy efficiency.

Programmes that have been operating over the past decade, and longer, have a history of proven savings in megawatts (MW), megawatt-hours (MWh), and therms, as well as on customer bills. These programmes show that energy efficiency can compare very favourably to supply-side options. This chapter summarises key findings from a portfolio-level review of many of the energy efficiency programmes that have been operating successfully for a number of years. It provides an overview of best practices in the following areas:

1. Political and human factors that have led to increased reliance on energy efficiency as a resource
2. Key considerations used in identifying target measures for energy efficiency programming in the near term and long term
3. Programme design and delivery strategies that can maximise programme impacts and increase cost-effectiveness
4. The role of monitoring and evaluation in ensuring that programme dollars are optimised and that energy efficiency in planning, design and implementation, and evaluation was derived from a review of energy efficiency programmes at the portfolio level across a range of policy models (e.g. public benefit charge administration, integrated resource planning)

A sustained history of successful energy efficiency programme implementation summarised and shared the following characteristics:

1. Significant investment in energy efficiency as a resource within their policy context
2. Development of cost-effective programmes that deliver results
3. Incorporation of programme design strategies that work to remove near-and long-term market barriers to investment in energy efficiency
4. Willingness to devote the necessary resources to make programmes successful

Most of the organisations reviewed also have conducted full-scale impact evaluations of their portfolio of energy efficiency investments within the last few years. The best practices gleaned from a review of these organisations can assist utilities, their commissions, state energy offices and other stakeholders in overcoming barriers to significant energy efficiency programming, and can help to begin tapping into energy efficiency as a valuable and clean resource to effectively meet future supply needs.

References

1. EnergyStar.gov. (2007). *Milestones: ENERGY STAR*, Retrieved March 1, 2008, from www.energystar.org
2. U.S. Environmental Protection Agency. *2006 Annual Report: ENERGY STAR and Other Climate Protection Partnerships*, Retrieved March 1, 2008, from www.EPA.org
3. EnergyStar.gov. *History: ENERGY STAR*. Retrieved March 1, 2008, from www.energystar.org
4. Green Energy Efficient Homes. Energy Efficient Dehumidifiers, www.energystar.org
5. EnergyStar.gov. *Business*. Retrieved March 2010, from www.energystar.gov
6. *Industries in Focus: ENERGY STAR*, Energystar.gov, 2009, Retrieved March 23, 2009, from http://www.energystar.gov/index.cfm?c=in_focus.bus_industries_focus#plant

9

Current Energy Leakages in Egyptian Buildings

9.1 Examples of Public and Residential Buildings

9.1.1 Building Blocks of a National Standards and Labelling Programme

A national standards and labelling programme is defined as a set of elements that ensure that energy efficiency standards and labelling efforts are effective, appropriate, strengthened over time and sustained [1–5]. The building blocks fall into two categories: technical/policy and process and should include the following.

9.1.2 Technical/Policy

1. Accredited testing facilities: Facilities should be internationally accredited and staffed with competent testing personnel; they should have the capacity to test models in a timely manner.

2. Appropriate testing procedures: Testing procedures are the methods by which the energy efficiency level of a product is deduced. The selected procedures should reasonably reflect the usage patterns and climate particular to a country. This builds consumer confidence that test results accurately reflect the energy usage he/she will experience.

3. Energy labels: Standards and labels can be established separately or as complementary programmes. Many types of labelling programmes exist.

4. Energy efficiency standards can be mandatory or voluntary and can be based either on maximum energy consumption or minimum energy efficiency.

5. An energy policy framework that is conducive to energy efficiency is critical to the longevity of a national standards programme. Supportive policies include government procurement requirements, voluntary programmes, incentives to manufacturers, consumer awareness campaigns, demand side management and integrated resource planning.

9.1.3 Proposed Process

1. Compliance with voluntary and mandatory standards and labelling requirements must be ensured through a credible enforcement scheme to guarantee programme effectiveness. Programme evaluation will inform necessary programme modifications, justify further activities and provide the documentation necessary to sustain the standards and labelling programmes over the long term.

2. The legislative process should ensure that standards and labels are periodically reviewed and raised ('ratcheted' upward) as the overall product efficiency on the market improves. The changes will mostly depend on the results of programme evaluation.

3. In the programme design and improvement process, input from all stakeholders (government, private companies, consumer associations etc.) should be considered. Cooperation among the stakeholders is the key to the success of programmes. However, the local and national governments must also hold their decisions final, after carefully considering all suggestions.

9.2 Measures and Assessments

9.2.1 Proposed Labels for Egypt: Major Appliances

The proposed energy labels are separated into at least four categories:

1. The appliance's details: Specific details of the model and its materials for each appliance.
2. Energy class: A colour code associated with a letter (from A to G) that gives an idea of the appliance's electrical consumption.
3. Consumption, efficiency, capacity and so forth: This section gives information according to appliance type.
4. Noise: The noise emitted by the appliance is described in decibels.

9.2.1.1 Refrigerators, Freezers and Combined Appliances

There should be a summary that indicates the energy efficiency; the index is calculated for each appliance according to its consumption and its compartments' volume taking into account the appliance type. The index is thus not calculated in kWh (Figure 9.1).

The label also contains:

1. The annual energy consumption in kWh per year
2. The capacity of fresh foods in litres for refrigerators and combined appliances
3. The capacity of frozen foods in litres for freezers and combined appliances
4. The noise in dB(A)

For cold appliance models (and this product alone) that are more economical than those of category A, categories A+ and A++ have been assigned.

9.2.1.2 Washing Machines and Tumble Dryers

For washing machines, the energy efficiency scale is calculated using a cotton cycle at 60°C with a maximum declared load. This load is typically 6 kg. The energy efficiency index is in kWh per kilograms of washing (Figure 9.2). The energy label also contains information on

1. Total consumption per cycle
2. Washing performance, with a class from A to G
3. Spin drying performance, with a class from A to G
4. Maximum spin speed
5. The total cotton capacity in kg
6. Water consumption per cycle in litres
7. Noise in the washing and spinning cycles in dB(A)

A++	A+	A	B	C	D	E	F	G
<30	<42	<55	<75	<90	<100	<110	<125	>125

FIGURE 9.1
Energy labels for refrigerators.

A	B	C	D	E	F	G
<0.19	<0.23	<0.27	<0.31	<0.35	<0.39	>0.39

FIGURE 9.2
Energy labels for washing machines.

For tumble dryers, the energy efficiency scale is calculated using the cotton drying cycle with a maximum declared load. The energy efficiency index is in kWh per kilograms of load. Different scales apply for condenser and vented dryers (Figure 9.3).

The label also contains:

1. The energy consumption per cycle
2. The total cotton capacity
3. Whether the unit is vented or condensing
4. Noise in dB(A)

For combined washer/dryers, the energy efficiency scale is calculated using the cotton drying cycle with a maximum declared load. The energy efficiency index is in kWh per kilograms of load. Different scales apply for condenser and vented dryers (Figure 9.4).

The label also contains:

1. The energy consumption per cycle (washing and drying)
2. The energy consumption per cycle (washing only)
3. Washing performance, with a class from A to G
4. The maximum spin speed
5. The total cotton capacity (washing and drying separately)

Condenser dryers

A	B	C	D	E	F	G
<0.55	<0.64	<0.73	<0.82	<0.91	<1.00	>1.00

Vented dryers

A	B	C	D	E	F	G
<0.51	<0.59	<0.67	<0.75	<0.83	<0.91	>0.91

FIGURE 9.3
Energy labels for dryers.

A	B	C	D	E	F	G
<0.68	<0.81	<0.93	<1.05	<1.17	<1.29	>1.29

FIGURE 9.4
Energy labels for combined washer/dryers.

6. Water consumption for a full load washed and dried; note that condenser dryers may use significant amounts of water on the drying cycle

7. Noise in dB (A) (separately for washing, spinning and drying)

9.2.1.3 Dishwashers

The energy efficiency is calculated according to the number of place settings. For the most common size of appliance (the 12 place setting machine), the following classes apply. The unit is expressed in kWh per 12 place settings (Figure 9.5).

The label also contains:

1. The energy consumption in kWh per cycle
2. The efficiency of the washing cycle, with a class from A to G
3. The efficiency of the drying cycle, with a class from A to G
4. The capacity as a number of place settings
5. The water consumption in litres per cycle
6. Noise in dB(A)

9.2.1.4 Air Conditioners

The directive applies only to units under 12 kW. On every label, you will find:

1. The model
2. The energy efficiency category from A to G
3. The annual energy consumption (full load at 500 hours per year)
4. The cooling output at full load in kW
5. The energy efficiency ratio in cooling mode at full load
6. The appliance type (cooling only, cooling/heating)
7. The cooling mode (air or water-cooled)
8. The noise rating in dB (where applicable)

A	B	C	D	E	F	G
<1.06	<1.25	<1.45	<1.65	<1.85	<2.05	>2.05

FIGURE 9.5
Energy labels for dishwashers.

	A	B	C	D	E	F	G
Cooling EER W/W	>3.2	3.0–3.2	2.8–3.0	2.6–2.8	2.4–2.6	2.2–2.4	<2.2
Heating COP W/W	>3.6	3.4–3.6	3.2–3.4	2.8–3.2	2.6–2.8	2.4–2.6	<2.4

FIGURE 9.6
Energy labels for air conditioners.

For air conditioners with heating capability (Figure 9.6), one can also find:

1. The heat output at full load in kW
2. The heating mode energy efficiency category

Note that there exist units with EER and COP greater than 5, so take a note of the actual number when it is A rated.

9.2.1.5 Light Bulbs

On every label, you will find:

1. The energy efficiency category from A to G
2. The luminous flux of the bulb in lumens
3. The electricity consumption of the lamp in watts
4. The average life length in hours

9.2.2 Concluding Remarks

It can be concluded that it is important to incorporate an energy performance directive as a standard in this region; such a goal will aid energy savings in large buildings and will set regulations for energy-efficient designs that are based on standard calculation methods. It is recommended to

1. Develop standardised tools for the calculation of the energy performance of buildings
2. Define system boundaries for the different building categories and different heating systems
3. Prepare models for expressing requirements on indoor air quality, thermal comfort in winter and, when appropriate, in summer, visual comfort and so on

4. Define comparable energy related key values (kWh/m^2, kWh per person, kWh per apartment, kWh per produced unit etc.) and develop a common procedure for an 'energy performance certificate'

5. Design, construct and operate a solar decathlon (building) that can meet rural and desert requirements and save diminishing fossil fuel sources

9.3 Laws, Codes and Standards

9.3.1 Building Mechanical Systems

Prescriptive Option Heating, Ventilating and Air Conditioning

Scope
This section specifies requirements for heating, ventilating and air-conditioning equipment efficiencies.

Compliance
The heating, ventilating and air conditioning shall comply with Section 5 of Egyptian Energy Efficiency Commercial Building Code (EEECBC) [6] with the following modifications and additions:

Minimum Equipment Efficiencies
Projects shall comply with one of the following:

1. Energy Code Baseline. Products with minimum efficiencies addressed in the HVAC Egyptian Code [6] provided that the *building project* contains:
   a. On-site renewable energy systems with 1.5 times the minimum capacity of that specified in Building Renewable Energy Systems [6].
   b. Peak load reduction systems with twice the peak load reduction specified in Peak Load Reduction/Load Factor. *Building projects* shall contain automatic systems such as demand limiting or load shifting to reduce Electric Peak demand of the building by not less than 5%. Standby power generation shall not be used to achieve the reduction in peak capacity.
2. Higher Efficiency. For those products where there is an Energy Saving programme, the minimum efficiency shall be the greater of the Energy Star requirements [5]. For other products, the equipment efficiency shall be a minimum of the values in Tables 9.1 through 9.7.

TABLE 9.1

Electrical Operated Unitary Air Conditioners and Condensing Units

Equipment Type	Size Category	Heating Section Type	Subcategory or Rating Conditions	Minimum Efficiency	Test Procedure
Air conditioners, air cooled	<19 kW	All	Split systems	4.10 SCOP 3.52 COP	ARI 210/240
			Single packaged	4.10 SCOP 3.52 COP	ARI 210/240
Through-the-wall air cooled	<9 kW	All	Split systems	3.52 SCOP	ARI 210/240
			Single packaged	3.52 SCOP	ARI 210/240
Small-duct high-velocity, air cooled	<19 kW	All	Split systems	2.93 SCOP	ARI 210/240
Air conditioners, air cooled	≥19 kW and <40 kW	Electric resistance (or none)	Split systems and single package	3.37 COP 3.52 ICOP	ARI 340/360
		All other	Split systems and single package	3.31 COP 3.46 ICOP	ARI 340/360
	≥40 kW and <70 kW	Electric resistance (or none)	Split systems and single package	3.37 COP 3.52 ICOP	ARI 340/360
		All other	Split systems and single package	3.31 COP 3.46 ICOP	ARI 340/360
	≥70 kW and <223 kW	Electric resistance (or none)	Split systems and single package	2.93 COP 3.08 ICOP	ARI 340/360
		All other	Split systems and single package	2.87 COP 3.02 ICOP	ARI 340/360
	≥223 kW	Electric resistance (or none)	Split systems and single package	2.84 COP 2.99 ICOP	ARI 340/360
		All other	Split systems and single package	2.78 COP 2.93 ICOP	ARI 340/360
Air conditioners, water and evaporative cooled	<19 kW	All	Split systems and single package	4.10 COP 4.19 ICOP	ARI 210/240
	≥19 kW and <40 kW	Electric resistance (or none)	Split systems and single package	4.10 COP 4.19 ICOP	ARI 340/360

TABLE 9.1 *(Continued)*

Electrical Operated Unitary Air Conditioners and Condensing Units

Equipment Type	Size Category	Heating Section Type	Subcategory or Rating Conditions	Minimum Efficiency	Test Procedure
		All other	Split systems and single package	4.04 COP 4.13 ICOP	ARI 340/360
	≥40 kW and <70 kW	Electric resistance (or none)	Split systems and single package	4.10 COP 4.19 ICOP	ARI 340/360
		All other	Split systems and single package	3.81 COP 4.13 ICOP	ARI 340/360
	≥70 kW	Electric resistance (or none)	Split systems and single package	4.10 COP 4.10 ICOP	ARI 340/360
		All other	Split systems and single package	3.81 COP 3.81 ICOP	ARI 340/360
Condensing units, air cooled	≥40 kW			[Not applicable match with indoor coil]	ARI 365
Condensing, water or evaporative cooled	≥40 kW			[Not applicable match with indoor coil]	ARI 365

9.3.2 Ventilation Controls for High-Occupancy Areas

Demand control ventilation (DCV) is required for *densely occupied spaces*. This requirement supersedes the occupant density threshold in Section 4 of EEECBC [6].

If CO_2 sensors are used as part of a DCV system, they shall be designed and installed as defined in the following section.

CO_2 Sensors

Spaces with CO_2 sensors or air-sampling probes leading to a central CO_2 monitoring station shall have one sensor or probe for each 10,000 ft^2 (1000 m^2) of floor space and shall be located in the room between 3 ft and 6 ft (1 m and 2 m) above the floor. CO_2 sensors must be accurate to ±50 ppm at 1000 ppm. For all spaces with CO_2 sensors, *target concentrations* shall be calculated for full and part load conditions (to include no less than 25%, 50%, 75% and 100%). Metabolic rate, occupancy, ceiling height and *outdoor air* CO_2

TABLE 9.2

Electrically Operated Unitary and Applied Heat Pumps, Minimum Efficiency Requirements

Equipment Type	Size Category	Heating Section Type	Subcategory or Rating Conditions	Minimum Efficiency	Test Procedure
Air conditioners, air cooled (cooling mode)	<19 kW	All	Split systems	4.10 SCOPC 3.52 COPC	ARI 210/240
			Single packaged	4.10 SCOPC 3.40 COPC	
Through-the-wall air cooled (cooling mode)	<9 kW	All	Split systems	3.52 SCOPC	
			Single packaged	3.52 SCOPC	
Small-duct high velocity, air cooled (cooling mode)	<9 kW	All	Split systems	2.93 SCOPC	
Air conditioners, air cooled (cooling mode)	≥19 kW and <40 kW	Electric resistance (or none)	Split systems and single package	3.31 COPC 3.46 ICOPC	ARI 340/360
		All other	Split systems and single package	3.25 COPC 3.40 ICOPC	
	≥40 kW and <70 kW	Electric resistance (or none)	Split systems and single package	3.32 COPC 3.46 ICOPC	
		All other	Split systems and single package	3.25 COPC 3.40 ICOPC	
	≥70 kW	Electric resistance (or none)	Split systems and single package	2.87 COPC 2.87 ICOPC	
		All other	Split systems and single package	2.81 COPC 2.81 ICOPC	

Equipment type	Capacity	Type	Conditions	Efficiency	Standard
Water source (cooling mode)	<5 kW	All	30°C Entering water	4.10 COPC	ISO-13256-1
	≥5 kW and <19 kW	All	30°C Entering water	4.10 COPC	
	>19 kW and <40 kW	All	30°C Entering water	4.10 COPC	
Groundwater source (cooling mode)	<40 kW	All	15°C Entering water	4.75 COPC	ARI 210/240
		All	25°C Entering water	13.4 COPC	
Air conditioners, air cooled (heating mode)	<19 kW	All	Split systems	2.49 SCOPH	
			Single packaged	2.34 SCOPH	
Through-the-wall air cooled (heating mode)	<9 kW	All	Split systems	2.17 SCOPH	
			Single packaged	2.17 SCOPH	
Small-duct high velocity, air cooled (heating mode)	<19 kW	All	Split systems	1.99 SCOPH	
Air cooled (heating mode)	≥19 kW and <40 kW (cooling capacity)		8°C DB/6 C WB outdoor air	3.3 COPH	ARI 340/360
			−8°C DB/−9 C WB outdoor Air	2.2 COPH	
	≥40 kW (cooling capacity)		8°C DB/6 C WB outdoor air	3.2 COPH	
			−8°C DB/−9 C WB outdoor air	2.0 COPH	
Water source (heating mode)	<40 kW (cooling capacity)		19°C Entering water	4.2 COPH	ISO-13256-1
Groundwater source (heating mode)	<19 kW (cooling capacity)		10°C Entering water	3.6 COPH	
			0°C Entering fluid	3.1 COPH	

TABLE 9.3

Water Chilling Packages – Minimum Efficiency Requirements

Equipment Type	Size Category	Units	Minimum Efficiency				Test Procedure
			Path A		Path B		
			Full Load	IPLV	Full Load	IPLV	
Air cooled chillers with condenser, electrically operated	<528 kW	COP	2.931	3.664	NA	NA	ARI 550/590
	≥528 kW	COP	2.931	3.737	NA	NA	
Air cooled without condenser, electrical operated	All capacities	COP	Condenser-less units shall be rated with matched condensers				ARI 550/590
Water-cooled, electrically operated, positive displacement (reciprocating)	All capacities	COP	Reciprocating units required to comply with water-cooled positive displacement requirements				ARI 550/590
Water-cooled electrically operated, positive displacement	<264 kW	COP	4.509	5.583	4.396	5.862	ARI 550/590
	≥264 kW and 528 kW	COP	4.538	5.719	4.452	6.002	
	≥528 kW and <1055 kW	COP	5.172	6.064	4.898	6.513	
	≥1055 kW	COP	5.673	6.513	5.504	7.178	
Water-cooled electrically operated, centrifugal	<528 kW	COP	5.547	5.901	5.504	7.816	ARI 550/590
	≥528 kW and <1055 kW	COP	5.547	5.901	5.504	7.816	
	≥1055 kW and <2110 kW	COP	6.106	6.406	5.862	8.792	
	≥2110 kW	COP	6.170	6.525	5.961	8.792	
Air cooled absorption single effect	All capacities	COP	0.600	NR	NA	NA	ARI 560
Water-cooled absorption single effect	All capacities	COP	0.700	NR	NA	NA	
Absorption double effect indirect-fired	All capacities	COP	1.000	1.050	NA	NA	
Absorption double effect direct fired	All capacities	COP	1.000	1.000	NA	NA	

Note: Where there is a Column A and Column B requirement, either column can be used for compliance, but both the full load and IPLV values shall be complied, with Path A intended for applications where significant operating time is expected at full load and design ambient and Path B intended for applications with significant operating time at part load. All Path B machines shall be equipped with demand-limiting capable controls; NA means that this category is not applicable and cannot be used for compliance; NR means that for this category there are no minimum requirements allowed to be used in heat recovery applications.

TABLE 9.4

Electrically Packaged Terminal Air Conditioners, Packaged Terminal Heat Pumps, Packaged Vertical Air Conditioners, Single Packaged Vertical Heat Pumps, Room Air Conditioners and Room Air Conditioner Heat Pumps – Minimum Efficiency Requirements

Equipment Type	Size Category (Input)	Subcategory or Rating Condition	Minimum Efficiency (SI)	Test Procedure
PTAC (cooling mode) new construction	<2.0 kW	35°C DB outdoor air	3.49 COPC	ARI 310/380
	≥2.0 kW and <2.9 kW	35°C DB outdoor air	3.31 COPC	
	≥2.9 kW and <3.8 kW	35°C DB outdoor air	3.14 COPC	
	≥3.8 kW	35°C DB outdoor air	3.48 COPC	
PTAC (cooling mode) replacement	<2.0 kW	35°C DB outdoor air	3.49 COPC	ARI 310/380
	≥2.0 kW and <2.9 kW	35°C DB outdoor air	3.31 COPC	
	≥2.9 kW and <3.8 kW	35°C DB outdoor air	3.14 COPC	
	≥3.8 kW	35°C DB outdoor air	3.48 COPC	
PTHP (cooling mode) new construction	<2.0 kW	35°C DB outdoor air	3.48 COPC	ARI 310/380
	≥2.0 kW and <2.9 kW	35°C DB outdoor air	3.48 COPC	
	≥2.9 kW and <3.8 kW	35°C DB outdoor air	3.48 COPC	
	≥3.8 kW	35°C DB outdoor air	3.48 COPC	
PTHP (heating mode) new construction	All capacities	35°C DB outdoor air	2.8 COPH	ARI 310/380
PTHP (cooling mode) replacement	<2.0 kW	35°C DB outdoor air	3.43 COPC	ARI 310/380
	≥2.0 kW and <2.9 kW	35°C DB outdoor air	3.25 COPC	
	≥2.9 kW and <3.8 kW	35°C DB outdoor air	3.08 COPC	
	≥3.8 kW	35°C DB outdoor air	2.73 COPC	
PTHP (heating mode) replacement	All capacities	35°C DB outdoor air	2.8 COPH	ARI 310/380
SPVAC (cooling mode)	<19 kW	35°C DB/23.9 C WB outdoor air	3.81 SCOPC	ARI 390
	>19 kW and <40 kW Btu/h	35°C DB/23.9 C WB outdoor air	3.37 COPC	
	>40 kW and <70 kW	35°C DB/23.9 C WB outdoor air	3.37 COPC	

Continued

TABLE 9.4 (Continued)

Electrically Packaged Terminal Air Conditioners, Packaged Terminal Heat Pumps, Packaged Vertical Air Conditioners, Single Packaged Vertical Heat Pumps, Room Air Conditioners and Room Air Conditioner Heat Pumps – Minimum Efficiency Requirements

Equipment Type	Size Category (Input)	Subcategory or Rating Condition	Minimum Efficiency (SI)	Test Procedure
SPVHP (cooling mode)	<19 kW	35°C DB/23.9 C WB outdoor air	3.81 SCOPC	ANSI/AHAM RAC-1
	>19 kW and 40 kW Btu/h	35°C DB/23.9 C WB outdoor air	3.37 COPC	
	>40 kW and <70 kW	35°C DB/23.9 C WB outdoor air	3.37 COPC	
SPVHP (heating mode)	<19 kW	8.3°C DB/6.1 C WB outdoor air	3.0 COPH	
	>19 kW and <40 kW Btu/h	8.3°C DB/6.1 C WB outdoor air	3.0 COPH	
	>40 kW and <70 kW	8.3°C DB/6.1 C WB outdoor air	2.9 COPH	
Room air conditioners, with louvred sides	<1.8 kW		3.14 SCOPC	
	>1.8 kW and <2.3 kW		3.14 COPC	
	>2.3 kW and <4.1 kW		3.17 COPC	
	>4.1 kW and <5.9 kW		3.14 COPC	
	>5.9 kW		2.73 COPC	
Room air conditioners, without louvred sides	<2.3 kW		2.90 COPC	
	>2.3 kW and <5.9 kW		2.73 COPC	
	>5.9 kW		2.73 COPC	
Room air conditioner heat pump with louvred sides	<5.9 kW		2.90 COPC	
	>5.9 kW		2.73 COPC	
Room air conditioner heat pump without louvred sides	<4.1 kW		2.73 COPC	
	>4.1 kW		2.58 COPC	
Room air conditioner, casement only	All capacities		2.81 COPC	
Room air conditioner, casement-slider	All capacities		3.05 COPC	

TABLE 9.5

Warm Air Furnaces and Combustion Warm Air Furnaces/Air-Conditioning Units, Warm Air Duct Furnaces and Unit Heaters

Equipment Type	Size Category (Input)[e]	Subcategory or Rating Condition	Test Procedure[b]	Minimum Efficiency
Warm air furnace, gas fired (outdoor installation)	<65.9 kW		DOE 10 CFR Part 430 or ANSI Z21.47	78% AFUE or 80% E_t[a]
	>65.9 kW	Maximum capacity[d]	ANSI Z21.47	80% E_c[c]
Warm air furnace, gas fired (indoor installation)	<65.9 kW		DOE 10 CFR Part 430 or ANSI Z21.47	90% AFUE or 92% E_t[a]
	>65.9 kW	Maximum capacity[d]	ANSI Z21.47	92% E_c[a]
Warm air furnace, oil fired (outdoor installation)	<65.9 kW		DOE 10 CFR Part 430 or UL 727	78% AFUE or 80% E_t[a]
	>65.9 kW	Maximum capacity[c]	UL 727	81% E_t[a]
Warm air furnace, oil fired (indoor Installation)	<65.9 kW		DOE 10 CFR Part 430 or UL 727	90% AFUE or 92% E_t[a]
	>65.9 kW	Maximum capacity[c]	UL 727	92% E_t[a]
Warm air duct furnaces, gas fired (outdoor installation)	All capacities	Maximum capacity[c]	ANSI Z83.9	80% E_c[a]
Warm air duct furnaces, gas fired (indoor installation)	All capacities	Maximum capacity[c]	ANSI Z83.9	90% E_c[a]
Warm air unit heaters, gas fired (indoor installation)	All capacities	Maximum capacity[c]	ANSI Z83.8	90% E_c[a]
Warm air unit heaters, oil fired (indoor installation)	All capacities	Maximum capacity[c]	UL 731	90% E_c[a]

Note: a. Efficiencies: Et = thermal efficiency. Ec = combustion efficiency. b. Units shall also include an interrupted or intermittent ignition device, have jacket lasses not exceeding 0.75% of the input rating and have either power venting or flue damper. c. A vent damper is an acceptable alternative to the fuel damper for those furnaces where combustion air is drawn from the *conditioned space*. d. Combustion units not covered by NAECA (3-phase power or cooling capacity greater than or equal to 19.0 kW) may comply with either rating. e. Minimum and maximum ratings as provided for and allowed by the unit's controls.

concentrations assumptions and *target concentrations* shall be shown in the design documents. *Outdoor air* CO_2 concentrations shall be determined by one of the following:

1. CO_2 concentration shall be assumed to be 400 ppm without any direct measurement.

2. CO_2 concentration shall be dynamically measured using a CO_2 sensor located near the position of the *outdoor air* intake.

TABLE 9.6

Gas and Oil Fired Boilers – Minimum Efficiency Requirements

Equipment Type	Subcategory or Rating Condition	Size Category (Input)	Efficiency	Test Procedure
Boilers, hot water	Gas fired	<87.9 kW	89% AFUE	10 CFR Part 430
		.87.9 kW and <732.7 kW	89% Et	10 CFR Part 431
		.732.7 kW	91% Ec	
	Oil fired	<87.9 kW	89% AFUE	10 CFR Part 430
		.87.9 kW and <732.7 kW	89% Et	10 CFR Part 431
		.732.7 kW	91% Ec	
Boilers, steam	Gas fired	<87.9 kW	75% AFUE	10 CFR Part 430
	Gas fired all, except natural draft	.87.9 kW and <732.7 kW	79% Et	10 CFR Part 431
		.732.7 kW	79% Et	
	Gas fired, natural draft	.87.9 kW and <732.7 kW	77% Et	
		.732.7 kW	77% Et	
	Oil fired	<87.9 kW	80% Et	10 CFR Part 430
		.87.9 kW and <732.7 kW	81% Et	10 CFR Part 431
		.732.7 kW	81% Et	

Ec = combustion efficiency (100% less flue losses).

Note: Units shall also include an interrupted or intermittent ignition device, have jacket losses not exceeding 0.75% of the input rating and have either power venting or flue damper. A vent damper is an acceptable alternative to the fuel damper for those furnaces where combustion air is drawn from the *conditioned space*.

      As of August 8, 2008, according to the Energy Policy Act of 2005, units shall also include an interrupted or intermittent ignition devices and have either power venting or automatic flue dampers. A vent damper is an acceptable alternative to a flue damper for those unit heaters where combustion air is drawn from the *conditioned space*.

Ducts and Plenum Leakage

For duct sealing, Seal Level A shall be used. This requirement supersedes the requirements in Section 5.3.4.6.2 of EEECBC.

Economisers

Systems shall have economisers as specified in Table 9.8 [6] and high-limit controls as specified in Table 9.9 [6]. Rooftop units with a capacity of less than 60,000 Btu/h (18 kW) shall have two stages of capacity control, with the first stage used for cooling with the economiser and the second stage to add mechanical cooling.

TABLE 9.7

Performance Requirements for Heat Rejection Equipment

Equipment Type	Total System Heat Rejection Capacity at Rated Conditions	Rating Standard	Rating Conditions (Annual Cooling Design, ASHRAE Standard 169, Table A1, Column 10c)	Performance Required	Full Load Maximum Approach C
Open loop propeller or axial fan cooling towers	All	CTI ATC-105 and CTI STD201	1.0% design evaporation WB temperature	>4.06 L/skW	3.3 C above 1.0% design evaporation WB temperature
Closed loop propeller or axial fan cooling towers	All	CTI ATC-105 and CTI STD201	1.0% design evaporation WB temperature	>1.52 L/skW	4.4 C above 1.0% design evaporation WB temperature
Open loop centrifugal fan cooling towers	All	CTI ATC-105 and CTI STD201	1.0% design evaporation WB temperature	>2.23 L/skW	3.3 C above 1.0% design evaporation WB temperature
Closed loop centrifugal fan cooling towers	All	CTI ATC-105 and CTI STD201	1.0% design evaporation WB temperature	>0.81 L/skW	4.4 C above 1.0% design evaporation WB temperature
Air-cooled condensers	All	ARI 460		[Not applicable, air cooled condenser shall be matched to the HVAC system and rated per tables C3]	

Note: These requirements apply to boilers with a rated input of 2344 kW or less that are not packaged boilers and to all packaged boilers. Minimum efficiency requirements for boilers cover all capacities of packaged boilers. E_c = thermal efficiency (100% less flue losses). See reference document for detailed information. E_t = thermal efficiency. See reference document for detailed information. Maximum capacity—minimum and maximum ratings as provided for and allowed by the unit's controls includes oil fired (residual). Systems shall be designed with lower operating return hot water temperatures (<55°C) and use hot water reset to take advantage of the higher efficiencies of condensing boilers. For purposes of this table, cooling tower performance is defined as the maximum flow rating of the tower divided by the fan nameplate rated motor power. For purposes of this table, air-cooled condenser performance is defined as the heat rejected from the refrigerant divided by the fan nameplate rated motor power. The approach is the design tower leaving water temperature—the building 1.0% annual cooling design evaporation WB temperature from ASHRAE standard.

TABLE 9.8

Minimum System Size for Which an Economiser Is Required

Climate Zones	Cooling Capacity for Which an Economiser Is Required
1A, 1B, 2A	No economiser requirement
2B, 3A, 3B, 3C, 4A, 4B, 4C, 5A, 5B, 5C, 6A, 6B, 7, 8	≥33,000 Btu/h (9.7 kW)[a]

Source: EEECBC EGYPTIAN Energy Efficiency Code, HBRC, 2012.

[a] Where economisers are required, the total capacity of all systems without economisers shall not exceed 480,000 Btu/h (140 kW) per building or 20% of the building's air economiser capacity, whichever is greater.

TABLE 9.9

High-Limit Shutoff Control Options for Air Economiser

Climate Zones	Allowable Control Types
1A, 2A, 3A, 4A	Differential enthalpy[a]
1B, 2B, 3B, 3C, 4B, 4C, 5A, 5B, 5C, 6A, 6B, 7, 8	Differential enthalpy or differential drybulb

Source: EEECBC EGYPTIAN Energy Efficiency Code, HBRC, 2012.

[a] Differential enthalpy is that between the return air and the outside air.

Zone Controls

Exception (a) to Section 6.5.2.1 of ASHRAE/IESNA Standard 90.1 [5] shall be replaced by the following: zones for which the volume of air that is reheated, re-cooled or mixed is not greater than the larger of (i) the *design outdoor airflow rate* for the zone, or (ii) 15% of the zone design peak supply rate.

Fan System Power Limitation Controls

Systems shall have fan power limitations 10% below limitations specified in Section 5.4.1 of EEECBC.

Exhaust Air Energy Recovery

The system shall comply with the energy recovery requirements within the threshold limits in Table 9.10. Where a single room or space is supplied by multiple units, the aggregate supply cfm (L/s) of those units shall be used in applying this requirement.

Energy recovery systems shall have a minimum of 60% recovery effectiveness. Sixty percent recovery effectiveness shall mean a change in the enthalpy of the *outdoor air* supply equal to the 60% of the difference between the *outdoor air* and return air enthalpies at design conditions. For equipment with airside economisers, provision shall be made for all outdoor and exhaust air to bypass the energy recovery device when not being used.

TABLE 9.10

Energy Recovery Requirement (SI)

	% Outside Air at Full Design Flow							
	≥10% and <20%	≥20% and <30%	≥30% and <40%	≥40% and <50%	≥50% and <60%	≥60% and <70%	≥70% and <80%	≥80%
Zone	Design Supply Fan Flow L/s							
3B, 3C, 4B, 4C, 5B	NR	NR	NR	NR	NR	NR	≥2360	≥2360
1B, 2B, 5C	NR	NR	NR	NR	≥12,271	≥5663	≥2360	≥1888
6B	NR	≥10,619	≥5191	≥2596	≥2124	≥1652	≥1180	≥708
1A, 2A, 3A, 4A, 5A, 6A	≥14,158	≥6135	≥2596	≥2124	≥1652	≥944	≥472	>0
7, 8	≥1888	≥1416	≥1180	≥472	>0	>0	>0	>0

Variable Speed Fan Control for Commercial Kitchen Hoods
In addition to the requirements in Section 6.5.7.1 of ASHRAE/IESNA Standard 90.1 [5], commercial kitchen Type I and Type II hood systems shall have variable speed control for exhaust and make-up air fans to reduce hood airflow rates at least 50% during those times when the cooking equipment is under no-load conditions.

Duct Insulation
Duct insulation shall comply with the minimum requirements in Tables 9.11 and 9.12 [6].

Pipe Insulation
Pipe insulation shall comply with the minimum requirements in Table 9.13 [5]. These requirements supersede the requirements in Table 6.8.3 of ASHRAE/IESNA Standard 90.1 [5].

Automatic Control of HVAC in Hotel/Motel Guest Rooms
A minimum of one of the following control technologies shall be required in hotel/motel guest rooms with over 50 guest rooms such that all the power to the lights and switched outlets in a hotel or motel guest room would be turned off when the occupant is not in the room and the space temperature would automatically setback (winter) or set up (summer) by no less than 20°C:

1. Controls that are activated by the room occupant via the primary room access method—key, card, deadbolt etc.

2. Occupancy sensor controls that are activated by the occupant's presence in the room.

TABLE 9.11

Minimum Duct Insulation R-Value, Cooling and Heating Only Supply Ducts and Return Ducts

Climate Zone	Exterior	Ventilated Attic	Unvented Attic above Insulated Ceiling	Unvented Attic with Roof Insulation	Unconditioned Space	Indirectly Conditioned Space	Buried
Heating Ducts Only							
1, 2	None	None	None	None	None	None	None
3	R-1.06	None	None	None	R-1.06	None	None
4	R-1.06	None	None	None	R-1.06	None	None
5	R-1.41	R-1.06	None	None	R 1.06	None	R-1.06
6	R-1.41	R-1.41	R-1.06	None	R 1.06	None	R-1.06
7	R-1.76	R-1.41	R-1.41	None	R-1.06	None	R-1.06
8	R-1.76	R-10	R-1.41	None	R-1.41	None	R-1.41
Cooling Only Ducts							
1	R-1.06	R-1.41	R-10	R-1.06	R-1.06	None	R-1.06
2	R-1.06	R-1.41	R-10	R-1.06	R-1.06	None	R-1.06
3	R-1.06	R-1.41	R-1.41	R-1.06	R-0.62	None	None
4	R-0.62	R-1.06	R-1.41	R-0.62	R-0.62	None	None
5, 6	R-0.62	R-0.62	R-1.06	R-0.62	R-0.62	None	None
7, 8	R-1.9	R-0.62	R-0.62	R-0.62	R-0.62	None	None
Return Ducts							
1–8	R-1.06	R-1.06	R-1.06	None	None	None	None

Source: EEECBC EGYPTIAN Energy Efficiency Code, HBRC, 2012.

Note: Insulation R-values, measured in m^2 k/kW, are for the insulation as installed and do not include film resistance. The required minimum thicknesses do not consider water vapour transmission and possible surface condensation. Where exterior walls are used as plenum walls, wall insulation shall be as required by the most restrictive condition of 6.4.4.2 or Section 5. Insulation resistance measured on a horizontal plane in accordance with ASTM C518 at a mean temperature of 23.8°C at the installed thickness. Includes crawl spaces, both ventilated and non-ventilated. Includes return air plenums with or without exposed roofs above.

TABLE 9.12

Minimum Duct Insulation R-Values, Combined Heating and Cooling Supply Ducts and Return Ducts

Climate Zone				Duct Location			
	Exterior	Ventilated Attic	Unvented Attic above Insulated Ceiling	Unvented Attic with Roof Insulation	Unconditioned Space	Indirectly Conditioned Space	Buried
Supply Ducts							
1	R-1.41	R-1.41	R-1.76	R-1.06	R-1.06	None	R-1.06
2	R-1.41	R-1.41	R-1.41	R-1.06	R-1.41	None	R-1.06
3	R-1.41	R-1.41	R-1.41	R-1.06	R-1.41	None	R-1.06
4	R-1.41	R-1.41	R-1.41	R-1.06	R-1.41	None	R-1.06
5	R-1.41	R-1.41	R-1.41	R-0.62	R-1.41	None	R-1.06
6	R-1.76	R-1.41	R-1.41	R-0.62	R-1.41	None	R-1.06
7	R-1.76	R-1.41	R-1.41	R-0.62	R-1.41	None	R-1.06
8	R-1.76	R-1.94	R-1.94	R-0.62	R-1.41	None	R-1.41
Return Ducts							
1–8	R-1.06	R-1.06	R-1.06	None	None	None	None

Source: EEECBC EGYPTIAN Energy Efficiency Code, HBRC, 2012.

Note: Insulation R-values, measured in $m^2 \, k/kW$, are for the insulation as installed and do not include film resistance. The required minimum thicknesses do not consider water vapour transmission and possible surface condensation. Where exterior walls are used as plenum walls, wall insulation shall be as required by the most restrictive condition of 6.4.4 2 or Section 5. Insulation resistance measured on a horizontal plane in accordance with ASTM C518 at a mean temperature of 23.8°C at the installed thickness. Includes crawl spaces, both ventilated and non-ventilated. Includes return air plenums with or without exposed roofs above.

TABLE 9.13

Minimum Pipe Thermal Insulation Thickness[a]

Fluid Design Operating Temp. Range, °C	Pipe Insulation Thickness						
	Insulation Conductivity		Nominal Pipe or Tube Size, mm				
	Conductivity, kW/(h-m²°K)	Mean Rating Temp, °C	<25	25 to <38	38 to <102	102 to <203	>203
Domestic and Service Hot Water Systems[a,b]							
40.6+	1.25–1.59	100	25	25	38	38	38
Heating Systems (Steam, Steam Condensate and Hot Water)[c,d]							
>176.7	1.82–1.93	121.1	76	89	89	114	114
121.7–176.7	1.65–1.82	200	51	76	89	89	89
93.9–121.1	1.53–1.70	150	51	51	63	63	63
60.6–93.3	1.42–1.65	125	38	38	38	51	51
40.6–60.0	1.25–1.59	100	25	25	38	38	38
Cooling Systems (Chilled Water, Brine and Refrigerant)[d]							
4.4–15.6	1.25–1.59	100	25	25	38	38	38
<4.4	1.25–1.59	100	25	38	38	38	51

Source: ASHRAE Standard 90.1—Energy Standard for Buildings Except Low-Rise Residential Buildings.

Note: a. For insulation outside the stated conductivity range, the minimum thickness (T) shall be determined as follows: T = r{(1 + t/r)K/k − 1}, where T = minimum insulation thickness (mm), r = actual outside radius of pipe (mm), t = insulation thickness listed in this table for applicable fluid temperature and pipe size, K = conductivity of alternate material at mean rating temperature indicated for the applicable fluid temperature (kW/h-m² K), and k = the upper value of the conductivity range listed in this table for the applicable fluid temperature. b. These thicknesses are based on energy efficiency considerations only. Additional insulation is sometimes required relative to safety issues/surface temperature. c. Piping insulation is not required between the control valve and coil on runouts when the control valve is located within 1.2 m of the coil and the pipe size is 25 mm or less. d. These thicknesses are based on energy efficiency considerations only. Issues such as water vapour permeability surface condensation sometimes require vapour retarders or additional insulation.

9.3.3 Building Service Water-Heating Systems

1. Service water heating: The *service water heating* shall comply with Section 6 of EEECBC [6] with the following modifications and additions:

2. Equipment efficiency: Equipment shall comply with the minimum efficiencies in Table 9.14 [6].

3. Service hot-water piping insulation: Pipe insulation shall comply with Table 9.13 above.

TABLE 9.14

Performance Requirements for Water-Heating Equipment

Equipment Type	Size Category (Input)	Subcategory or Rating Condition	Performance Required	Test Procedure
Electric water heaters	12 kW	Resistance ≥75.7L	0.93–0.00132V EF	DOE 10 CFR Part 430
	>12 kW	Resistance ≥75.7L	20 + 35 V.5 SL, Btu/h	ANSI Z21.10.3
	24 Amps and ≤250 Volts	Heat Pump	0.93–0.00132V EF	DOE 10 CFR Part 430
Gas storage water heaters	≤75,000 Btu/h	Resistance ≥75.7L	0.62–0.0019V EF	DOE 10 CFR Part 430
	>75,000 Btu/h	<310.1 (kW/L)	80% Et (Q/800 + 110 V.5) SL, Btu/h	ANSI Z21.10.3
Gas instantaneous water heaters	>50,000 Btu/hand <2000,000 Btu/h	≥310.1 kWQ/L and <7.56 L	0.62–0.0019V EF	DOE 10 CFR Part 430
	≥200,000 Btu/h	≥310.1 kWQ/L and <37.8 L	80% Et	ANSI Z21.10.3
	≥200,000 Btu/h	≥310.1 kWQ/L and <37.8 L	80% Et (Q/800 + 110 V.5) SL, Btu/h	
Oil storage water heaters	≤105,000 Btu/h	Resistance ≥75.7L	0.59–0.0019V EF	DOE 10 CFR Part 430
	>105,000 Btu/h	<310.1 (kW/L)	78% Et (Q/800 + 110 V.5) SL, Btu/h	ANSI Z21.10.3
Oil instantaneous water heaters	≤210,000 Btu/h	≥310.1 kWQ/L and <7.56 L	0.59–0.0019V EF	DOE 10 CFR Part 430
	>210,000 Btu/h	≥310.1 kWQ/L and <37.8 L	80% Et	ANSI Z21.10.3
	>210,000 Btu/h	≥310.1 kWQ/L and <37.8 L	78% Et (Q/800 + 110 V.5) SL, Btu/h	
Hot water supply boilers, gas and oil	300,000 Btu/hand <12,500,000 Btu/h	≥310.1 kWQ/L and <37.8 L	80% Et	ANSI Z21.10.3
Hot water supply boilers, gas		≥310.1 kWQ/L and <37.8 L	80% Et (Q/800 + 110 V.5) SL, Btu/h	
Hot water supply boilers, oil		≥310.1 kWQ/L and <37.8 L	78% Et (Q/800 + 110 V.5) SL, Btu/h	
Pool heaters oil and gas	All		78% Et	ASHRAE 146
Heat pump pool heaters	All		4.0 COP	ASHRAE 146
Unfired storage tanks	All		R-12.5	(None)

Source: EEECBC EGYPTIAN Energy Efficiency Code, HBRC, 2012.

4. Insulation for spa pools: Pools heated to more than 90°F (32°C) shall have side and bottom surfaces insulated on the exterior with a minimum insulation value of R-12 (R-2.1).

9.3.4 Energy-Saving Equipment

The following equipment within the scope of the applicable energy-saving programme shall comply with the equivalent criteria required to achieve the energy-saving label if installed prior to the issuance of the certificate of occupancy:

1. Appliances
   a. Clothes washers
   b. Dishwashers
   c. Refrigerators and freezers
   d. Room air conditioners
   e. Room air cleaners
   f. Water coolers
2. Heating and cooling
   a. Boilers
   b. Central air conditioners
   c. Ceiling fans
   d. Ventilating fans
3. Electronics
   a. Cordless phones
   b. Combination units (TV/VCR/DVD)
   c. DVD products
   d. Audio
   e. Televisions
4. Office Equipment
   a. Computers
   b. Copiers
   c. Fax machines
   d. Laptops
   e. Mailing machines
   f. Monitors
   g. Multifunction devices (printer/fax/scanner)

    h. Printers

    i. Scanners

5. Water heaters

6. Lighting

    a. Compact fluorescent light bulbs (CFLs)

7. Commercial food service

    a. Commercial fryers

    b. Commercial hot food holding cabinets

    c. Commercial solid door refrigerators and freezers

    d. Commercial steam cookers

    e. Commercial ice machines

    f. Commercial dishwashers

8. Other products

    a. Battery charging systems

    b. External power adapters

    c. Vending machines

9.4 Energy Auditing

9.4.1 General

An energy audit is an inspection, survey and analysis [7] of energy flows for energy conservation in a building, process or system to reduce the amount of energy input into the system without negatively affecting the output(s). When the object of study is an occupied building, reducing energy consumption while maintaining or improving human comfort, health and safety is of primary concern. Beyond identifying the sources of energy use, an energy audit seeks to prioritise the energy uses according to the greatest to least cost-effective opportunities for energy savings.

A home energy audit is a service where the energy efficiency of a house [7] is evaluated by a person using professional equipment (such as blower doors and infrared cameras) with the aim of suggesting the best ways to improve energy efficiency in heating and cooling the house.

An energy audit of a home may involve recording information about various characteristics of the building envelope including the walls, ceilings, floors, doors, windows and skylights. For each of these components, the area and resistance to heat flow (R-value) is measured or estimated. The leakage rate or infiltration of air through the building envelope is of concern; both of which are strongly affected by window construction and quality of door seals such as weather stripping. The goal of this exercise is to quantify the building's overall thermal performance. The audit may also assess the efficiency, physical condition and programming of mechanical systems such as the heating, ventilation, air-conditioning equipment and thermostat.

A home energy audit may include a written report estimating energy use given local climate criteria, thermostat settings, roof overhang and solar orientation. This could show energy use for a given time period, say a year, and the impact of any suggested improvements per year. The accuracy of energy estimates are greatly improved when the homeowner's billing history is available showing the quantities of electricity, natural gas, fuel oil or other energy sources consumed over a one- or two-year period. Some of the greatest effects on energy use are user behaviour, climate and age of the home. An energy audit may therefore include an interview of the homeowners to understand their patterns of use over time. The energy billing history from the local utility company can be calibrated using heating degree day and cooling degree day data obtained from recent, local weather data in combination with the thermal energy model of the building. Advances in computer-based thermal modelling can take into account many variables affecting energy use.

A home energy audit is often used to identify cost-effective ways to improve the comfort and efficiency of buildings. In addition, homes may qualify for energy efficiency grants from the central government.

Recently, improvements in smartphone technology have enabled homeowners to perform relatively sophisticated energy audits of their own homes. This technique has been identified as a method to accelerate energy efficiency improvements.

The term energy audit is commonly used to describe a broad spectrum of energy studies ranging from a quick walk-through of a facility to identify major problem areas to a comprehensive analysis of the implications of alternative energy efficiency measures sufficient to satisfy the financial criteria of sophisticated investors. Numerous audit procedures have been developed for non-residential (tertiary) buildings (ASHRAE [1–5]). An audit is required to identify the most efficient and cost-effective energy conservation opportunities (ECOs) or measures (ECMs). ECOs (or ECMs) can consist of more efficient use of energy or of partial or global replacement of the existing installation.

When looking at existing audit methodologies, the main issues of an audit process are

1. The analysis of building and utility data, including study of the installed equipment and analysis of energy bills
2. The survey of the real operating conditions
3. The understanding of the building behaviour and of the interactions with weather, occupancy and operating schedules

4. The selection and the evaluation of ECMs

5. The estimation of energy-saving potential

6. The identification of customer concerns and needs

Common types/levels of energy audits are distinguished here, although the actual tasks performed and level of effort may vary with the consultant providing services under these broad headings. The only way to ensure that a proposed audit will meet your specific needs is to spell out those requirements in a detailed scope of the work. Taking the time to prepare a formal solicitation will also assure the building owner of receiving competitive and comparable proposals. Generally, four levels of analysis can be outlined (ASHRAE):

1. Level 0—Benchmarking: This first analysis consists of a preliminary whole building energy use (WBEU) analysis based on historic utility use and costs and on the comparison of the performance of the buildings to those of similar buildings. This benchmarking of the studied installation allows the determination of whether further analysis is required.

2. Level I—Walk-through audit: A preliminary analysis is made to assess building energy efficiency to identify not only simple and low-cost improvements but also a list of ECMs or ECOs to orient the future detailed audit. This inspection is based on visual verifications, study of installed equipment and operating data and detailed analysis of recorded energy consumption collected during the benchmarking phase.

3. Level II—Detailed/general energy audit: Based on the results of the pre-audit, this type of energy audit consists of an energy use survey to provide a comprehensive analysis of the studied installation, a more detailed analysis of the facility, a breakdown of the energy use and a first quantitative evaluation of the ECOs/ECMs selected to correct the defects or to improve the existing installation. This level of analysis can involve advanced on-site measurements and sophisticated computer-based simulation tools to evaluate precisely the selected energy retrofits.

4. Level III—Investment-grade audit: This is a detailed analysis of capital-intensive modifications focusing on potential costly ECOs requiring rigorous engineering study.

9.4.2 Benchmarking

The impossibility of describing all possible situations that might be encountered during an audit means that it is necessary to find a way of describing what constitutes good, average and bad energy performance across a range of situations [7–8]. The aim of benchmarking is to answer this question. Benchmarking mainly consists of comparing the measured consumption with reference consumption of other similar buildings, or it is generated by simulation tools to identify excessive or unacceptable running costs. As mentioned before, benchmarking is also necessary to identify buildings presenting interesting energy-saving potential. An important issue in benchmarking is the use of performance indexes to characterise the building.

These indexes can be

1. Comfort indexes, comparing the actual comfort conditions to the comfort requirements
2. Energy indexes, consisting of energy demands divided by heated/conditioned area, allowing comparison with reference values of indexes coming from regulations or from similar buildings
3. Energy demands, directly compared to 'reference' energy demands generated by means of simulation tools

9.4.2.1 Walk-Through or Preliminary Audit

The preliminary audit (alternatively called a simple audit, screening audit or walk-through audit) is the simplest and quickest type of audit. It involves minimal interviews with site operating personnel, a brief review of facility utility bills and other operating data and a walk-through of the facility to become familiar with the building operation and to identify any glaring areas of energy waste or inefficiency.

Typically, only major problem areas will be covered during this type of audit. Corrective measures are briefly described, and quick estimates of implementation cost, potential operating cost savings and simple payback periods are provided. A list of ECMs or ECOs requiring further consideration is also provided. This level of detail, although not sufficient for reaching a final decision on implementing a proposed measure, is adequate to prioritise energy efficiency projects and to determine the need for a more detailed audit.

9.4.2.2 General Audit

The general audit (alternatively called a mini-audit, site energy audit, detailed energy audit or a complete site energy audit) expands on the preliminary audit described earlier by collecting more detailed information about facility operation and by performing a more detailed evaluation of ECMs. Utility bills are collected for a 12- to 36-month period to allow the auditor to evaluate the facility's energy demand rate structures and energy usage profiles. If interval meter data are available, the detailed energy profiles that such data make possible will typically be analysed for signs of energy waste. Additional metering of specific energy-consuming systems is often performed to supplement utility data. In-depth interviews with facility operating personnel are conducted to provide a better understanding of major energy-consuming systems and to gain insight into short and longer term energy consumption patterns. This type of audit will be able to identify all ECMs appropriate for the facility, given its operating parameters. A detailed financial analysis is performed for each measure based on detailed implementation cost estimates, site-specific operating cost savings and the customer's investment criteria. Sufficient detail is provided to justify project implementation. The evolution of cloud-based energy auditing software platforms is enabling the managers of commercial buildings to collaborate with general and specialty trades contractors in performing general and energy system-specific audits [6–8]. The benefit of software-enabled collaboration is the ability to identify the full range of energy efficiency options that may be applicable to the specific building under study with 'live time' cost and benefit estimates supplied by local contractors.

9.4.2.3 Investment-Grade Audit

In most corporate settings, upgrades to a facility's energy infrastructure must compete for capital funding with non-energy-related investments. Both energy and non-energy investments are rated on a single set of financial criteria that generally stress the expected return on investment. The projected operating savings from the implementation of energy projects must be developed such that they provide a high level of confidence. In fact, investors often demand guaranteed savings. The investment-grade audit expands on the detailed audit described earlier and relies on a complete engineering study in order to detail technical and economical issues necessary to justify the investment related to the transformations.

9.4.2.3.1 Simulation-Based Energy Audit Procedure for Non-Residential Buildings
The following procedure proposes to make an intensive use of modern building energy systems tools at each step of the audit process, from benchmarking to detailed audit and financial study:

1. Benchmarking stage: Although normalisation is required to allow comparison between data recorded on the studied installation and reference values deduced from case studies or statistics, the use of simulation models to perform a code-compliant simulation of the installation under study allows direct assessment of the studied installation, without any normalisation needed. Indeed, applying a simulation-based benchmarking tool allows an individual normalisation and avoids size and climate normalisation.

2. Preliminary audit stage: Global monthly consumptions are generally insufficient to allow an accurate understanding of the building's behaviour. Even if the analysis of the energy bills does not allow identifying with accuracy the different energy consumers present in the facility, the consumption records can be used to calibrate building and system simulation models. To assess the existing system and to simulate correctly the building's thermal behaviour, the simulation model has to be calibrated on the studied installation. The iterations needed to perform the calibration of the model can also be fully integrated in the audit process and can help in identifying required measurements and critical issues [7].

3. Detailed audit stage: At this stage, on-site measurements, sub-metering and monitoring data are used to refine the calibration of the Building Energy Systems (BES) tool. Extensive attention is given to understanding not only the operating characteristics of all energy-consuming systems but also situations that cause load profile variations on short and longer term bases (e.g. daily, weekly, monthly, annually). When the calibration criteria is satisfied, the savings related to the selected ECOs/ECMs can be quantified [7–9].

4. Investment-grade audit stage: At this stage, the results provided by the calibrated BES tool can be used to assess the selected ECOs/ECMs and to orient the detailed engineering study.

5. Infrared thermography audit: The advent of high-resolution thermography has enabled inspectors to identify potential issues within the building envelope by taking a thermal image of the various surfaces of a building. For purposes of an energy audit, the thermographer will analyse the patterns within the surface temperatures to identify heat transfer through convection, radiation or conduction. It is important to note that the thermography ONLY identifies SURFACE temperatures, and analysis must be applied to determine the reasons for the patterns within the surface temperatures. Thermal analysis of a home generally costs between US$300 and US$600. For those who cannot afford a thermal inspection, it is possible to get a general feel for the heat loss with a non-contact infrared thermometer and several sheets of reflective insulation. The method involves measuring the temperatures on the inside surfaces of several exterior walls to establish baseline temperatures. After this, reflective barrier insulation is taped securely to the walls in 8-foot (2.4 m) by 1.5-foot (0.46 m) strips, and the temperatures are measured in the centre of the insulated areas at 1-hour intervals for 12 hours. (The reflective barrier is pulled away from the wall to measure the temperature in the centre of the area that it has covered.) The best manner in which to do this is when the temperature differential (Delta T) between the inside and outside of the structure

is at least 40 degrees. A well-insulated wall will commonly change approximately one degree per hour if the difference between external and internal temperatures is an average of 40 degrees. A poorly insulated wall can drop as much as 10 degrees in an hour.

9.4.2.4 Pollution Audits

With increases in carbon dioxide emissions or other greenhouse gases, pollution audits are now a prominent factor in most energy audits. Implementing energy-efficient technologies helps prevent utility-generated pollution.

Online pollution and emission calculators can help approximate the emissions of other prominent air pollutants in addition to carbon dioxide.

Pollution audits generally take electricity and heating fuel consumption numbers over a two-year period and provide approximations for carbon dioxide, VOCs, nitrous oxides, carbon monoxide, sulphur dioxide, mercury, cadmium, lead, mercury compounds, cadmium compounds and lead compounds.

References

1. ASHRAE Guideline 1.1—HVAC and R Technical Requirements for The Commissioning Process, Latest Edition, 2013.
2. ASHRAE Standard 52.2—Method of Testing General Ventilation Air-Cleaning Devices for Removal Efficiency by Particle Size, Latest Edition 2010.
3. ASHRAE Standard 55—Thermal Environmental Conditions for Human Occupancy, 2013.
4. ASHRAE Standard 62.1—Ventilation for Acceptable Indoor Air Quality—Sets the minimum acceptable ventilation requirements, 2010.
5. ASHRAE Standard 90.1—Energy Standard for Buildings Except Low-Rise Residential Buildings.
6. EEECBC EGYPTIAN Energy Efficiency Code, HBRC, 2012.
7. ISO 16818: Energy Definitions, 2008.
8. Khalil, E. E. (2012). International focus on emerging technologies & opportunities, *ASME Congress*, Paper Number IMECE2012-94127.
9. Khalil, E. E. (2012). An international outlook of innovative design of low carbon buildings, *Keynote Paper at 3rd International Symposium on Low Carbon Buildings (ISLCB)*, Ningbo-China, October 2012.

10

Future Prospects

10.1 Continuous Urbanisation and Climate Change

Climate change is the current focus of urban debates because it is becoming a reality and not just anticipated at some future date. Urban areas are major contributors to the phenomenon as well as victims of it. Major urbanisation activities take place in cities of the developing world, where informalisation is synonymous with urbanisation.

This current process of urbanisation is seen by many scientists as one of the reasons for climate change; urban areas are where most of the CO_2 emissions from the built environment are produced and where the greatest consumers live, resulting in very high ecological footprints. The recent report, *Working Group I Contribution to the IPCC Fifth Assessment Report: Climate Change 2013*, states that 'It is extremely likely that human influence has been the dominant cause of the observed warming since the mid-20th century' [1]. Of course, cities are suffering to different extents from the consequences of climate change, whether from unprecedented heat waves, cold weather, storms, tsunamis and so on. Thus, the issue of investigating the mutual relationship between cities and climate change becomes more vital.

It is widely known that energy use is the major contributor to climate change. Cities are the major consumers of energy, which makes the issue of energy efficiency vital to limiting, if not reversing, the effects of climate change.

10.2 Energy Efficiency and Urban Expansion

Currently, urban expansion takes one of three forms: new cities being built, cites expanding through new suburbs or cities growing through informal expansion. All these forms pose a great threat to the city in the urban century. However, they require different approaches as to whether to confine or to improve their use of different resources, with energy as the prime resource. There is a need for a shift away from the car-based urban planning that dominated the last century to urban planning on a more human scale. New cities should be planned as self-sufficient units that offer a variety of housing and job opportunities to maximise walkability and minimise car dependency. Expanding cities should follow a more polycentric trend, where new districts are created and not exclusive residential areas. This can help create more jobs and, again, minimise commuting, thus decreasing energy consumption. Moreover, a shift toward public transit systems is essential to enhance the efficient use of available energy and resources. This is a notion well addressed by transit-oriented development.

Regarding informal expansion, the people's way of developing is 'smart'; however, their settlements need a more comprehensive approach to ensure the provision of needs (such as housing and services) in a more environmentally responsive pattern. This must be the basis for projects whether they are upgrades or redevelopment. During redevelopment, it is important to maintain high densities, walkability, mixed uses, mixed housing and to integrate with nearby open spaces. Providing access to safe water, sanitation, secure tenure, durable housing and sufficient living area should not divert projects from attending to other aspects of sustainability.

Some studies have shown that informal areas consume less energy due to their compact design. However, this compactness and extensive use of bricks, concrete and asphalt increases the effect of the urban heat island in the area, thus reducing comfort in outdoor spaces and increasing thermal loads on buildings. There is still a need to further investigate this issue in order to identify adequate intervention to improve environmental performance and to reduce energy use.

10.3 Integrating Renewable Energies into Urban Planning

Globally, there is a growing concern with reaching the 20% renewable energy targets of 2020* and going beyond that to 30% or to doubling the current share of renewables in the energy mix by 2030[†] and afterward. This target should be integrated in the urban planning system whether relating to decentralised production (by various users) or centralised production in power plants. In both cases, this could include regulations to facilitate installation of solar and wind energy production equipment. Currently, in some countries, urban regulations might restrict the installation of photovoltaic (PV) modules on

---

* The EU target for 2020 is 20% of the energy demand from renewable energy as stated in European Union [2].
† Within the upcoming Sustainable Development Goals (post 2015), Goal 7 states: Ensure access to affordable, sustainable and reliable modern energy services for all. Relevant targets are: 7.2 double the share of renewable energy in the global energy mix by 2030; 7.3 double the global rate of improvement in energy efficiency by 2030 as stated in Ref. [3].

roofs or on facades. This is especially apparent in new suburbs or in historic districts where regulations regarding heights and facade colours and materials might need to change to accommodate the new installations.

On the urban scale, a new paradigm should govern how we plan our cities, orient our buildings and plot our roads. Different variations are needed to improve the environmental performance of the city and to reduce thermal loads on buildings, thereby reducing the need for future cooling/heating and the accompanying consumption of energy. Other issues are how to smartly plan the locations for needed renewable power plants and how to connect them efficiently to different buildings. Planning new cities should have these criteria embedded in the infrastructure of the city and not simply as later add-ons. This approach is extensively adopted in city of Masdar in the United Arab Emirates.

10.4 Energy-Efficient Building Envelopes

The building envelope, also known as the building shell, fabric or enclosure, is the boundary between the conditioned interior of a building and the outdoors. The energy performance of building envelope components, including external walls, floors, roofs, ceilings, windows and doors, is critical in determining how much energy is required for heating and cooling. The building envelope's impact on energy consumption should not be underestimated: globally, space heating and cooling account for more than one-third of all energy consumed in buildings, rising to as much as 50% in cold climates and more than 60% in the residential subsector in cold climate countries.

Overall, one can conclude that buildings are responsible for more than 35% of global energy consumption. It is very alarming that, although whole-building approaches are ideal, every day building envelope components are upgraded or replaced using technologies that are less efficient than the best options available. These advanced options, which are among the primary focus of this book, are needed not only to support whole-building approaches but also to improve the energy efficiency of individual components.

Building envelopes are expected to save almost 6 exajoules (EJ) in 2050 under the two-degree Celsius scenario (equivalent to the current energy consumption of the United Kingdom). The goal is to change the way of constructing building to save energy: an inevitable solution.

The main issues are as follows:

1. The construction of new buildings should offer the best opportunity to deploy passive heating and cooling designs, which make use of energy-efficient building materials to minimise energy required for heating and cooling.

2. Transforming typical building renovation to make way for deep reductions in energy consumption, known as deep renovation, should be a high priority.

3. Well-planned building envelope design improvements can enhance occupant comfort and the quality of life of millions of citizens, while offering significant non-energy benefits such as reduced health care costs and reduced mortality of 'at risk' populations.

4. Airtight sealed buildings—restricting the passage of air through the building envelope—are key ways of increasing energy efficiency during new construction and deep renovation. Energy audits, such as the energy performance certificates that are mandatory in the European Union, should include regular, validated testing of air leakage (e.g. at least every ten years) in addition to worldwide enforcement.

5. New office buildings should be fitted with integrated facade systems that optimise daylight while minimising energy requirements for heating, cooling, artificial lighting and peak electricity use.

6. It is vital to increase global collaboration on developing more affordable zero-energy buildings, especially in cold climates.

Recommendations would foster the following technologies that would lead to greater returns on investment:

1. Highly insulated windows
2. Advanced, high-performance, 'thin' insulation
3. Less labour-intensive air sealing and lower-cost validation testing
4. Lower-cost automated dynamic shading and glazing
5. More durable and lower-cost reflective roof materials and reflective coatings

10.5 Providing Integrated Solutions for Low-Income Areas Constituting Most Urbanisation Activities

Climate change is gradually beginning to dominate urban debates of the 21st century. It is widely discussed in the domain of developed countries. Moreover, when it is rarely discussed in developing countries, the debates focus on the adaptation and mitigation necessary for risk zones. However, debates concerning improving the performance of cities in these countries have received minimum attention up till now. The challenge that lies in these cities is that they are mostly informal, thus requiring different approaches and solutions. It is estimated that more than 60% of Cairo is made up of informal areas, with more areas added daily. This poses a number of concerns, such as how these areas contribute to the growing phenomenon of climate change, how these areas will be affected, and most importantly, how these areas can assist in improving the environmental conditions and reversing the effects of climate change.

This area is yet to be developed because, currently, it lacks adequate research. There are some initiatives worldwide to address this issue; however, they are still individual instances that have not yet formed policies. Growing recognition of this issue was addressed by the Sustainable Development Goals (SDGs). Target 7.6 requires 'By 2030, expand and upgrade as appropriate infrastructure for supply, transmission, and distribution of modern and renewable energy services in rural and urban areas, including with a view to doubling primary energy supply per capita for Least Developing Countries (LDCs') [3].

10.6 Awareness Regarding Climate Change, Energy Crisis and Resource-Efficient Urbanism

Climate change is a driving issue in current debates regarding sustainable urban development. It is closely linked to issues of resilience and to the ability to withstand shocks, especially regarding weather catastrophes. It is apparent that informal areas are more efficient than other areas; however, they contribute to increasing the causes of climate change by their continuous encroachment on agricultural land or on risk zones and by their emission of CO_2 without any absorption. It is vital to note that these areas are efficient in reusing their products; however, garbage separation and recycling has not received much attention. This is an issue that accentuates their contribution to climate change.

A number of interventions could help improve the performance of urban areas (of which informal areas comprise a majority in the developing countries' cities) in combating climate change. However, a vital issue is awareness regarding the importance of climate change, its magnitude and its effects.

1. Many efforts should be directed toward raising awareness about climate change plans for both municipalities and residents.

2. Enhancing cyclability and promoting cycling through awareness campaigns would highlight its benefits and its trendiness worldwide.

3. The use of solar energy should be encouraged, especially because many cities in the developing world have very good solar exposure, in many applications (such as water heaters or street lighting). This could be done through a government initiative to subsidise the initial cost or to offer other appropriate incentives.

4. Another effort would be to promote interventions that can help combat climate change and reduce heat island effects and that can contribute to better use of energy, such as roof top planting (green roofs).

5. In arid zones, awareness could be promoted regarding the use of the ecological landscape to ensure shading in arid zones and to reduce the use of water.

6. Encouraging garbage separation and recycling would preserve resources and maximise the benefit from latent energy used during initial manufacturing.

10.7 Bottom-Up Approaches versus Top-Down Interventions

Awareness and initiatives related to energy efficiency and the use of renewables are not limited to national governments and to international organisations and policies.* They are gaining wide recognition on the local levels too. There are numerous efforts in the field, including local government; civic organisations such as community-based organisations (CBOs), nongovernmental organisations (NGOs), or customers' rights protection organisations; research institutes; the private sector and its Small and Medium Enterprises (SMEs); or individual interest groups. Some of these initiatives are emerging in Cairo, the second most polluted city of the world. One of these initiatives, the Participatory Development Programme in Urban Areas, has been undertaken by the German Technical Cooperation (GIZ). The overall objective of this initiative is to promote resilient communities and their efforts to adapt to the changing climate conditions and to increase information levels among public administration and civil society regarding the consequences of climate change, particularly in informal areas.[†,‡]

---

* Current international efforts include: The Millennium Development Goals and the Post 2015 Sustainable Development Goals promoted by the UN. On the regional level, there is the Arab strategy for Housing and Sustainable Urban Development promoted by the Arab League, and the EU Growth Strategy 2020.

† Main objectives of this programme are (1) development of a participatory and community-based adaptation strategy for informal areas in collaboration with Egyptian partners; (2) implementation of small-scale adaptation measures in informal areas in collaboration with residents and civil society in order to enhance urban resilience; and (3) awareness raising and capacity development among local residents, civil society, ministries, public agencies and local administration on climate change adaptation and urban resilience.

‡ Other partners are the Egyptian Environmental Affairs Agency (EEAA), the Egyptian Meteorological Authority (EMA), the Governorates of Cairo and Giza, the United Nations International Strategy for Disaster Reduction (UN-ISDR) and the Information and Decision Support Center (IDSC), www.egypt-urban.net; Contact: Carl Philipp Schuck, carl-philipp. schuck@giz.de

Another initiative is the Ecocitizen World Map project comprising a public–private partnership that is led by the U.S. NGO Ecocity Builders and joined by Esri, the Association of American Geographers, Eye on Earth (a partnership of United Nations Environment Programme (UNEP) and Abu Dhabi Environmental Data Initiative), Cairo University, Mundiapolis University, University of California at Berkeley and a number of local NGOs and community partners. The project, which started in 2013, provides tools and training to citizens, public officials and others who want to ensure a more sustainable and resilient urban environment through more informed decision making. The project is intended to offer training that will provide hands-on experience with spatial data collection using smartphones and tablets as well as low-cost open source tools for civic science in order to increase public participation and legibility around the collection of high-quality, environmentally and socially relevant data. Citizen reports will be layered over authoritative data for visualisation and analysis through an online map portal to provide decision makers at every level with additional insights and bottom-up motivation for more sustainable living, planning and urban design. The project is based on broader scientific research defined by the International Ecocity Framework and Standards (IEFS) initiative currently under development by the U.S. NGO Ecocity Builders and its international expert advisory, and in consultation with the United Nations International Strategy for Disaster Reduction (UNISDR's) 'Making Cities Resilient Campaign' and UN-Habitat's World Urban Campaign and Resilient Cities Profiling Program [4,5].

In addition, awareness campaigns led by local CBOs are numerous, whether promoting energy-efficient behaviour or encouraging the use of more energy-efficient appliances. Moreover, there is a growing market for Small and Medium Enterprises (SMEs) in the field of renewable energy, especially solar water heaters and Photo Voltaic (PV) modules that contribute to decentralised energy production.

10.8 New Business Models of Green Economy that Support Energy-Efficient Urban Development

There is still much to be studied to identify possible tailored solutions and how to implement them in a financially viable manner to ensure their continuity. Currently, there is a need to support the emerging businesses within the sought green economy. A set of enabling legislations and incentives is vital to promote new business models that can effectively contribute to the conversion into an energy-efficient urban development. There is a great need for the contribution and participation of various organisations, international and local, governmental and non-governmental, to move effectively toward achieving energy efficiency and renewable energy targets. This is mandatory in order to ensure a better environment for current and future generations.

References

1. IPCC, 2013: Summary for Policymakers. In: Climate Change 2013: The Physical Science Basis. Contribution of Working Group I to the Fifth Assessment Report of the Intergovernmental Panel on Climate Change [Stocker, T.F., D. Qin, G.-K. Plattner, M. Tignor, S.K. Allen, J. Boschung, A. Nauels, Y. Xia, V. Bex and P.M. Midgley (eds.)]. Cambridge University Press, Cambridge, United Kingdom and New York, NY, USA.
2. European Union. (2013). *The European Union Explained: Europe 2020: Europe's Growth Strategy*, European Commission Directorate-General for Communication Publications, Brussels.
3. UN General Assembly. (2014). Report of the Open Working Group of the General Assembly on Sustainable Development Goals, Sixty – eighth session, A/68/970, Retrieved August 15, 2014, from http://undocs.org/A/68/970.
4. The Project. (2014). Ecocitizen World Map Project, Retrieved June 2, 2014, from http://ecocitizenworldmap.org/about/the-project/.
5. IEFS Indicator Development. (2014), International Ecocity Framework and Standards, Retrieved June 2, 2014, from http://www.ecocitystandards.org/ecocity-level-1-conditions/iefs-indicator-development/.

Bibliography

80plus.org. (2007). The 80 plus program: About, Retrieved March 3, 2007, from http://www.plugloadsolutions.com/About.

Andrey, J., Johnson, L. C., Moos, M., and Whitfield, J. (2008). Does design matter? The ecological footprint as a planning tool at the local level, *Journal of Urban Design*, Vol. 11, No. 2, pp. 195–224.

Arandel, C. and El Batran, M. (1997). *The Informal Housing Development Process in Egypt*, Centre Nacional de la Recherche Scientifique, Bordeaux.

ASHRAE. (2011). *Applications*, ASHRAE, Atlanta, GA.

ASHRAE. (2013). *Handbook, Fundamentals*, ASHRAE, Atlanta, GA.

ASHRAE Standards 55-2010, ASHRAE, Atlanta.

Associated Consultants. (2008). *Strategic Planning for Ashmun City, Menoufia, Egypt*, Ministry of Housing, Utilities and Urban Communities and General Organization of Physical Planning (GOPP), Cairo, Egypt.

Associated Consultants and Little, A. D. (2010). *10th of Ramadan Strategic Plan*, General Organisation of Physical Planning and Ministry of Housing, Utilities and Urban Development, Cairo.

Banister, D. (1992). Energy use, transport and settlement patterns, In M. Breheny, ed., *Sustainable Development and Urban Form*, Pion Ltd, London, pp. 160–181.

Belzer, R. (2008). *Energy Star appliances: EPA's savings calculator exaggerates savings, Regulatory Economics*, Retrieved March 5, 2008, from www.neutralsource.org/archives/595.

Ben-Hamouche, M. (2008). Climate, cities and sustainability in the Arabian region: Compactness as a new paradigm in urban design and planning, *Archnet-IJAR, International Journal of Architectural Research*, Vol. 2, No. 2, pp. 196–208.

Berridge Lewinberg Greenberg, Ltd. (1991a). *Guidelines for the Reurbanisation of Metropolitan Toronto*, Municipality of Metropolitan Toronto Corporate Printing Services, Toronto, Canada.

Berridge Lewinberg Greenberg, Ltd. (1991b). *Study of the Reurbanisation of Metropolitan Toronto*, Municipality of Metropolitan Toronto Corporate Printing Services, Toronto, Canada.

Blomquist, G., Berger, M., and Hoehn, J. (1988). New estimates of the quality of life in urban areas. *American Economic Review*, Vol. 78, No. 1, pp. 89–107.

Boarnet, M. G. and Crane, R. (2001). *Travel by Design. The Influence of Urban Form on Travel*, Oxford University Press, New York.

Breheny, M. (1992a). The contradictions of compact city: A review, In M. Breheny, ed., *Sustainable Development and Urban Form*, Pion Ltd, London, pp. 138–159.

Breheny, M., ed. (1992b). *Sustainable Development and Urban Form*, Pion Ltd, London.

Breheny, M. (1996). Centrists, decentrists and compromisers: Views on the future of urban form, In M. Jenks, E. Burton, and K. Williams, eds., *The Compact City: A Sustainable Urban Form?* E & FN Spon, London, pp. 13–35.

Brindle, R. E. (1994). Lies, damned lies and 'automobile dependence'—Some hyperbolic reflections, *Australian Transport Research Forum*, Vol. 94, pp. 117–131.

Buxton, M. (2000). Energy, transport and urban form in Australia, In M. Jenks, E. Burton, and K. Williams, eds., *Achieving Sustainable Urban Form*, E & FN Spon, London, pp. 54–63.

Cairo Governorate; Housing and Utilities Sector. (2009). *Zeinhom Gardens Housing Project*, Cairo Governorate, Cairo.

California Sustainability Alliance Energy Star Televisions, Retrieved July 24, 2010, from http://sustainca.org/techshowcase/est/overview.

Calthorpe, P. (1992). The pedestrian pocket: New strategies for suburban growth, In B. Walter et al., eds., *Sustainable Cities: Concepts and Strategies for Eco-City Development*, Eco-Home Media, Los Angeles, CA, pp. 27–35.

Calthorpe, P. (1993). *The Next American Metropolis*, Princeton Architectural Press, New York.

Chambers, N., Simmons, C., and Wackernagel, M. (2000). *Sharing Nature's Interest: Ecological Footprints as an Indicator of Sustainability*, Earthscan, London.

Chen, Q., Cheung, G., Hu, Y., Shen, Q., Tang, B., and Yeung, S. (2009). A system dynamics model for the sustainable land use planning and development, *Habitat International*, Vol. 33, pp. 15–25.

Cho, Y. and Awbi, H. B. (2002). Effect of heat source location in a room on the ventilation performance, *ROOMVENT 2002*, pp. 445–448.

Churchman, A. (1993). A differentiated perspective on urban quality of life: Women, children and the elderly, In M. Bonnes, ed., *Perception and Evaluation of Environmental Quality*, UNESCO Programme on Man and Biosphere, Rome, pp. 165–178.

Churchman, A. (1999). Disentangling the concept of density, *Journal of Planning Literature*, Vol. 13, No. 4, pp. 389–411.

Commission of the European Communities (CEC). (1990). *Green Paper on the Urban Environment*, European Commission, Brussels.

Committee for The Development of Environmental Performance Assessment Tool for Cities and Institute for Building Environment and Energy Conservation. (2011). *Overview of CASBEE for Cities*, Japan Sustainable Building Consortium (JSBC), Tokyo.

Congress for the New Urbanism, Natural Resources Defense Council, and U.S. Green Building Council. (2012). *LEED 2009 for Neighborhood Development Rating System*, U.S. Green Building Council, Washington, DC.

Corgnati, S. P., Fracastoro, G. V., and Perino, M. 2002. Influence of cooling strategies on the air flow pattern in an office with mixing ventilation, *ROOMVENT 2002*, pp. 165–168.

Costanza, R., Fisher, B., Ali, S., Beer, C., Bond, L., Boumans, R., Danigelis, N. L., et al. (2008). An integrative approach to quality of life measurement, research and policy, *Surveys and Perspectives Integrating Environment and Society*, Vol. 1, pp. 11–15.

Crilly, M. (1999). Novocastrian urbanism, *Urban Design Quarterly*, Vol. 72, p. 10.

Criteria for Rating Building Energy Performance, Energystar.gov. Retrieved March 23, 2009, from http://www.energystar.gov/index.cfm?c = eligibility.bus_portfoliomanager_eligibility

Cummins, R. (1997). *Comprehensive quality of life scale—Adult*, School of Psychology, Deakin University, Melbourne.

Cummins, R. A., Eckersley, R., Pallant, J., Vugt, J. V., and Misajon, R. (2003). Developing a national index of subjective wellbeing: The Australian unity wellbeing index. *Social Indicators Research*, Vol. 64, pp. 159–190.

DEFRA. (2002). *Energy Resources. Sustainable Development and Environment*, Department of Environment, Food and Rural Affairs (DEFRA), Doncaster, UK.

Deutsch, L., Jansson, A., Troell, M., Ronnback, P., Folke, C., and Kautsky, N. (2000). The ecological footprint: Communicating human dependence on nature's work, *Ecological Economics*, Vol. 32, No. 3, pp. 351–356.

Devuyst, D., Hens, L., and De Lannoy, W. (2003). *How Green is the City? Sustainability Assessment and the Management of Urban Environments*, Columbia University Press, New York.

Diener, E. and Suh, E. (1997). Measuring quality of life: Economic, social, and subjective indicators, *Social Indicators Research*, Vol. 40, pp. 189–216.

Dramstad, W. E., Olson, J., and Forman, R. (1996). *Landscape Ecology Principles in Landscape Architecture and Land Use Planning*, Island Press, Washington, DC.

Draper, D. (1998). *Our Environment, A Canadian Perspective*, Thomson Nelson, Toronto.

Drewnowski, J. (1974). *On Measuring and Planning the Quality of Live*, Mouton, published for the Institute of Social Studies, The Hague.

Eckersley, R. (1998). Perspectives on progress: Economic growth, quality of life and ecological sustainability, In R. Eckersley, ed., *Measuring Progress: Is Life Getting Better?* CSIRO Publishing, Melbourne, pp. 3–34.

Ecocity Standards. (2011). *International Ecocity Framework and Standards*, Retrieved January 15, 2014, from http://www.ecocitystandards.org/

Edahiro, J. (2009). *Conceptual Basis of The Movement to Create and Propagate 'Eco-Model Cities' Initiatives of the Japanese Government*, JFS Newsletter No. 78, Retrieved June 8, 2011, from http://www.japanfs.org/en/mailmagazine/newsletter/pages/028824.html

Elkin, T., Mclaren, D., and Hillman, M. (1991). *Reviving the City: Towards Sustainable Urban Development*, Friends of the Earth, London.

ENERGY STAR. (2009). *Qualified Homes: ENERGY STAR*, Energystar.gov. Retrieved March 23, 2009, from http://www.energystar.gov/homes

EnergyStar.gov. *History: ENERGY STAR*, Retrieved March 1, 2008, from www.energystar.gov

EnergyStar.gov. *Learn More about Energy Guide: Energy Star*, Retrieved March 1, 2008, from www.energystar.gov

EnergyStar.gov. (2007). *Milestones: ENERGY STAR*, Retrieved March 1, 2008, from www.energystar.gov

EnergyStar.gov. *Room Air Conditioners Key Product Criteria*, Retrieved July 17, 2008, Energystar.gov; Retrieved March 23, 2009, from http://www.energystar.gov/index.cfm?c = roomac.pr_crit_room_ac

Energy Stars May Not be All They Say They Are, Housingzone.com. Retrieved March 23, 2009, http://www.housingzone.com/articleXml/LN888056763.html

Environmental News Service, Energy Star Climate Change Claims Misleading, Audit Finds, Washington, DC., December 31, 2008.

Enwicht, D. (1992). *Towards an Eco-city: Calming the Traffic*, Envirobook, Sydney.

EPD. (2006). *European Energy Performance Code*, CEN, Brussels.

Erikson, R. (1993). Descriptions of inequality: The Swedish approach to welfare research, In M. Nussbaum and A. Sen, eds., *The Quality of Life*, Clarendon Press, Oxford, pp. 67–87.

Erikson, R. and Uusitalo, H. (1987). The Scandinavian approach to welfare research, In R. Erikson, E. Hansen, S. Ringen, and H. Uusitalo, eds., *The Scandinavian Model: Welfare States and Welfare Research*, M. E. Sharpe, New York, pp. 177–193.

European Commission Energy Labeling, EPBD, Brussels, 2010.

European Union. (2013). *The European Union Explained: Europe 2020: Europe's Growth Strategy*, European Commission Directorate-General for Communication Publications, Brussels.

Evill, B. (1995). Population, urban density, and fuel use: Eliminating spurious correlation, *Urban Policy and Research*, Vol. 13, No. 1, pp. 29–36.

Ewing, B., Moore, D., Goldfinger, S., Oursler, A., Reed, A., and Wackernagel, M. (2010). *The Ecological Footprint Atlas 2010*, Global Footprint Network, Oakland, CA.

Expert group meeting on cities in climate change, *Urban Environment Newsletter*, February 2008, Urban Environment Section, UN-HABITAT, Nairobi, Kenya, in collaboration with UNEP, DTIE, Urban Environment Unit, Retrieved February 15, 2009, from http://www.unhabitat.org/scp

Farr, D. (2008). *Sustainable Urbanism: Urban Design with Nature*, John Wiley, Hoboken, NJ.

Federal Register/Vol. 72, No. 245/Friday, December 21, 2007/Rules and Regulations 72565, Retrieved www.gpo.gov/

Forum for The Future and General Electric GE. (2010). *The Sustainability Cities Index 2010*, Forum for The Future and GE, London.

Frey, H. (1999). *Designing the City: Towards a More Sustainable Urban Form*, Spon Press, London.

Gabriel, S. A., Mattey, J., and Wascher, W. L. (2003). Compensating differentials and evolution in the quality-of-life among U.S. states, *Regional Science and Urban Economics*, Vol. 33, No. 5, pp. 619–649.

Gabriel, S. A. and Rosenthal, S. S. (2004). Quality of the business environment versus quality of life: Do firms and households like the same cities? *The Review of Economics and Statistics*, Vol. 86, No. 1, 438–444.

General Organisation of Physical Planning, Giza Governorate, GTZ, and Ministry of Housing, Utilities and Urban Development. (2004). *Boulaq Al-Dakrour District Guide Plan 2004–2017*, GTZ, Cairo.

General Organization of Physical Planning (GOPP) and United Nations Development Program (UNDP). (2007). *Improving Living Condition within Informal Settlements through Adopting Participatory Planning: General Framework for Upgrading and Controling Informal Areas*, GOPP, Cairo.

General Organization for Physical Planning (GOPP) and United Nations Human Settlements Project (UN-Habitat). (2008). *Terms of Reference for Preparing Strategic Urban Planning for Small Cities in Egypt*, UN-Habitat, Cairo, Egypt.

Gerardet, H. (2008). *Cities People Planet: Urban Development and Climate Change*, 2nd ed., John Wiley, West Sussex, England.

Girardet, H. (2004). *Urban Planning and Sustainable Energy: Theory and Practice*, Forum Barcelona, Retrieved May 12, 2009, from http://www.barcelona2004.org/eng/banco_del_conocimiento/dialogos/fichac390.html

GlobeScan and MRC McLean Hazel. (2007). *Megacity Challenges: A stakeholder perspective*, Siemens AG, Munich.

Gordon, D. and Tamminga, K. (2002). Large-scale traditional neighbourhood development and pre-emptive ecosystem planning: The markham experience, 1989–2001, *Journal of Urban Design*, Vol. 7, No. 3, pp. 321–340.

Gordon, P. and Richardson, H. (1997). Are compact cities a desirable planning goal? *Journal of the American Planning Association*, Vol. 63, No. 1, pp. 95–106.

Gordon, P. and Richardson, H. (1998). Farmland preservation and ecological footprints: A critique, *Planning and Markets*, Vol. 1, No. 1, pp. 1–7.

Gordon, P. and Richardson, H. W. (1989). Gasoline consumption and cities—A reply, *Journal of the American Planning Association*, Vol. 55, No. 3, pp. 342–345.

Gowling, D. and Penny, L. (1988). Urbanisation, planning and administration in the London region: Processes of a metropolitan culture, In H. van der Cammen, ed., *Four Metropolises in Western Europe*, Van Gorcum, Maastricht, The Netherlands, pp. 5–59.

Green City Index. (2012). Siemens, Retrieved May 7, 2012, from http://www.siemens.com/entry/cc/en/greencityindex.htm

GTZ. (2007). *Participatory Development in Urban Areas Program, Boulaq AlDakrour District*, GTZ, Cairo.

Gyourko, J. and Tracy, J. (1991). The structure of local public finance and the quality of life, *Journal of Political Economy*, Vol. 99, No. 4, pp. 774–806.

Handy, S. (1996). Understanding the link between urban form and nonwork travel behavior, *Journal of Planning Education and Research*, Vol. 15, No. 3, pp. 183–198.

Hasic, T. (2000). A sustainable urban matrix: Achieving sustainable urban form in residential buildings, In M. Jenks, E. Burton, and K. Williams, eds., *Achieving Sustainable Urban Form*, E & FN Spon, London, pp. 329–336.

Herendeen, R. (2000). Ecological footprint is a vivid indicator of indirect effects, *Ecological Economics*, Vol. 32, No. 3, pp. 357–358.

Hitchcock, J. (1994). *A Primer on The Use of Density in Land Use Planning, Papers on Planning and Design no. 41*, Program in Planning, University of Toronto, Toronto, Canada.

Holden, E. (2004). Ecological footprints and sustainable urban form, *Journal of Housing and the Built Environment*, Vol. 19, No. 1, pp. 91–109.

Holden, E. and Norland, I. T. (2005). Three challenges for the compact city as a sustainable urban form: Household consumption of energy and transport in eight residential areas in the greater Oslo region, *Urban Studies*, Vol. 42, No. 12, pp. 2145–2166.

Holmberg, J., Lundqvist, U., Robert, K. H. and Wackernagel, M. (1999). The ecological footprint from a systems perspective of sustainability, *International Journal of Sustainable Development and World Ecology*, Vol. 6, No. 1, pp. 17–33.

Holmberg, S. and Einberg, G. (2002). Flow behavior in a ventilated room—Measurements and simulations, *ROOMVENT 2002*, pp. 197–200.

Hough, M. (1995). *Cities and Natural Process*, Routledge, London.

Høyer, K. G. and Holden, E. (2003). Household consumption and ecological footprints in Norway: Does urban form matter? *Journal of Consumer Policy*, Vol. 26, pp. 327–349.

Hruska, J. (2009). *Sony LCD Exceeds Energy Star Power Draw 75% of Time*, Arstechnica.com, Retrieved March 23, 2009, from http://arstechnica.com/gadgets/news/2009/02/sony-lcd-exceeds-energy-star-power-draw-75-of-time.ars

Hruska, J. (2010). *Fake Products and Companies Certified by Energy Star, Popular Mechanics*. Retrieved March, 2010, from http://www.popularmechanics.com/technology/industry/4350335.html

http://www.energystar.gov/index.cfm?c=business.bus_bldgs

http://www.energystar.gov/index.cfm?c=evaluate_performance.bus_portfoliomanager

http://www.iclei.org/our-activities/our-agendas/low-carbon-city/gcc.html

ICLEI Local Governments for Sustainability is the world's leading association of more than 1000 metropolises, cities, urban regions and towns representing over 660 million people in 86 countries. ICLEI promotes local action for global sustainability and supports cities to become sustainable, resilient, resource efficient, biodiverse, low carbon; to build a smart infrastructure; and to develop an inclusive, green urban economy with the ultimate aim of achieving healthy and happy communities, Retrieved January 30, 2014, from http://www.iclei.org

Imperato, I. and Ruster, J. (2003). *Slum Upgrading and Participation: Lessons from Latin America*, The World Bank, Washington, DC.

Industries in Focus: ENERGY STAR, Energystar.gov, 2009, Retrieved March 23, 2009, from http://www.energystar.gov/index.cfm?c = in_focus.bus_industries_focus#plant

IPCC, 2013: Summary for Policymakers. In Climate Change 2013: The Physical Science Basis. Contribution of Working Group I to the Fifth Assessment Report of the Intergovernmental Panel on Climate Change [T. F. Stocker, D. Qin, G.-K. Plattner, M. Tignor, S. K. Allen, J. Boschung, A. Nauels, Y. Xia, V. Bex, and P. M. Midgley, eds.]. Cambridge University Press, Cambridge, United Kingdom and New York, NY, USA.

ISO 13790, *Thermal Performance of Buildings—Calculation of Energy Use for Space Heating*, ISO, 2010.

ISO publications ISO 23045_2009, ISO, Geneva, Switzerland, January 2009.

Jacobs, J. (1961). *The Death and Life of Great American Cities: The Failure of Town Planning*, Random House, New York.

Jacobsen, T. S., Hansen, R., Mathiesen, E., Nielsen, P. V., and Topp, C. (2002). Design method and evaluation of thermal comfort for mixing and displacement ventilation, *ROOMVENT 2002*, pp. 209–212.

Jenks, M., Burton, E., and Williams, K., eds. (1996). *The Compact City: A Sustainable Urban Form?* E & FN Spon, London.

Jenks, M., Burton, E., and Williams, K., eds. (2000a). *Achieving Sustainable Urban Form*, E & FN Spon, London.

Jenks, M., Burton, E., and Williams, K. (2000b). Achieving sustainable urban form: An introduction, In M. Jenks, E. Burton, and K. Williams, eds., *Achieving Sustainable Urban Form*, E & FN Spon, London, pp. 1–6.

Jenks, M., Burton, E., and Williams, K. (2000c). Achieving sustainable urban form: Conclusions, In M. Jenks, E. Burton, and K. Williams, eds., *Achieving Sustainable Urban Form*, E & FN Spon, London, pp. 347–355.

Kahn, M. (1995). A revealed preference approach to ranking city quality of life, *Journal of Urban Economics*, Vol. 38, pp. 221–235.

Kameel, R. (2002). *Computer Aided Design of Flow Regimes in Air-Conditioned Operating Theatres*, PhD thesis, Cairo University.

Kameel, R. and Khalil, E. E. (2001). Numerical computations of the fluid flow and heat transfer in air-conditioned spaces, NHTC2001-20084, *35th National Heat Transfer Conference*, Anaheim, CA.

Kameel, R. and Khalil, E. E. (2002a). Prediction of flow, turbulence, heat-transfer and air humidity patterns in operating theatres, *ROOMVENT 2002*, pp. 69–72.

Kameel, R. and Khalil, E. E. (2002b). Prediction of turbulence behaviour using k-ε model in operating theatres, *ROOMVENT 2002*, pp. 73–76.

Kameel, R. and Khalil, E. E. (2003). Energy efficiency, merits, and advantages of various air-conditioning system in commercial buildings, in Egypt, *Proceedings of ASHRAE-RAL, Paper Ral.3-5*, Cairo.

Kennedy, M. (1995). *Ekologisk stadsplanering I Europa, In Den miljövänliga staden— en Utopi?* [Ecological urban planning in Europe], Rapport fra n en seminarserie, Miljöprosjekt Sankt Jörgen, Gøteborg.

Khalil, E. E. (2003). HVAC in energy efficiency building code, Egypt, *Proceedings of ASHRAE-RAL, Paper Ral.3-6*, Cairo.

Khalil, E. E. (2004a). Energy efficiency in air conditioned buildings: An overview, *Proceedings of 6th JIMEC*, Amman.

Khalil, E. E. (2004b). Energy efficiency in air conditioned buildings in Egypt, *Proceedings of 6th JIMEC*, Amman.

Khalil, E. E. (2005). Energy performance of buildings directive in Egypt: A new direction, *HBRC Journal*, Vol. 1, pp. 15–27.

Khalil, E. E. (2006). Energy performance of commercial buildings in Egypt: A new direction, *Proceedings, Energy2030*, Abu Dhabi, November 2006.

Khalil, E. E. (2008a). Air conditioning and refrigeration code for energy-efficient buildings in the Arab world, *Journal of Kuwait Society of Engineers*, Vol. 100, pp. 94–95.

Khalil, E. E. (2008b). Arab-air conditioning and refrigeration code for energy-efficient buildings, *Arab Construction World*, Vol. 28, No. 8, pp. 24–26.

Khalil, E. E. (2009). Thermal management in hospitals: Comfort, air quality and energy utilization, *Proceedings ASHRAE, RAL*, Kuwait, October 2009.

Khalil, E. E. (2010). Ten tips for energy-efficient air-conditioned buildings, *Proceedings of DETC, ASME International Design Engineering Technology Conference, DETC2010-28348*, Montreal, Canada, August 2010.

Khalil, E. E. (2011). Ventilation of the archaeological tombs of the valley of kings, Luxor, Egypt, *Proceedings of International Conference on Air-Conditioning & Refrigeration 2011, ICACR2011, ICACR2011-00102*, Korea.

Khalil, E. E. (2012a). An international outlook of innovative design of low carbon buildings, *Keynote paper at 3rd International symposium on low carbon buildings (ISLCB)*, Ningbo, China, October 2012.

Khalil, E. E. (2012b). Energy performance of commercial buildings: A new direction, *Proceedings, ASME IMECE*, November 2012.

Khalil, E. E. (2012c). International focus on emerging technologies & opportunities, *ASME Congress, Paper Number IMECE2012-94127*.

Khalil, E. E., Medhat, A. A., Morkos, S. M., and Salem, M. Y. (2012). Applications of neural network in predicting energy consumption in administrative buildings, *AIAA Annual Meeting*, January 2012.

Khalil, H. (2009a). Energy efficiency strategies in urban planning of cites, *7th International Energy Conversion Engineering Conference IECEC 2009*. AIAA, Orlando.

Khalil, H. (2009b). Energy efficiency strategies in urban planning of cities, *45th AIAA/ASME/SAE/ASEE Joint Propulsion Conference & Exhibit and 7th Annual International Energy Conversion Engineering Conference*, 2–5 August, 2009, Denver, Co, USA, paper No. AIAA 2009-4622.

Khalil, H. (2010). New urbanism, smart growth and informal areas: A quest for sustainability, In Steffen Lehmann, Husam AlWaer, and Jamal Al-Qawasmi, eds., *Sustainable Architecture & Urban Development*, CSAAR, Amman, pp. 137–156.

Khalil, H. (2012a). Affordable housing: Quantifying the concept in the Egyptian context, *Journal of Engineering and Applied Science*, Vol. 59, No. 2, pp. 129–148.

Khalil, H. (2012b). Enhancing quality of life through strategic urban planning, *Sustianable Cities and Society*, Vol. 5, pp. 77–86.

Khalil, H. (2012c). Sustainable urbanism: Theories and green rating systems, *48th AIAA/ASME/SAE/ASEE Joint Propulsion Conference & Exhibit and 10th Annual International Energy Conversion Engineering Conference*, 30 July–1 August, 2012, Atlanta, GA, USA, paper No. AIAA 2012-4248.

Kosonen, R. (2002). Displacement ventilation for room air moisture control in hot and humid climate, *ROOMVENT 2002*, pp. 241–244.

Land, K. C. (1996). Social indicators and the quality of life: Where do we stand in the mid-1990s? *SINET*, Vol. 45, pp. 5–8.

Lehman and Associates. (1995). *Urban Density Study: General Report*, Office for the Greater Toronto Area, Toronto, Canada.

Leite, B. C. C. and Tribess, A. (2002). Analysis of under floor air distribution system: Thermal comfort and energy consumption, *ROOMVENT 2002*, pp. 245–248.

Leite, C. and Tello, R. (2011). *Megacity Sustainability Indicators*. Retrieved April 15, 2012, from http://www.stuchileite.com

Leung, H. L. (2003). *Land Use Planning Made Plain*, 2nd ed., University of Toronto Press, Toronto.

Liu, Y. and Moser, A. (2002). Airborne particle concentration control for an operating room, *ROOMVENT 2002*, pp. 229–232.

Llewelyn-Davies, South Bank University, Environment Trust Associates, and London Planning Advisory Committee. (1994). *The Quality of London's Residential Environment*, London Planning Advisory Committee, London.

Madanipour, A. (1996). *Design of Urban Space*, John Wiley, New York.

McGranahan, G. (2005). An overview of urban environmental burdens at three scales: Intra-urban, urban-regional, and global, special feature on the environmentally sustainable city, *International Review for Environmental Strategies*, Vol. 5, No. 2, pp. 335–356.

McGregor, J. A., Camfield, L., and Woodcock, A. (2009). Needs, wants and goals: Wellbeing, quality of life and public policy, *Applied Research Quality Life*, Vol. 4, pp. 135–154.

McHarg, I. (1969). *Design with Nature*, The Natural History Press, Garden City, NY.

McLaren, D. (1992). Compact or dispersed? Dilution is no solution, *Built Environment*, Vol. 18, No. 4, pp. 268–284.

McMullen, D. (2000). High densities—Key to meeting housing needs, *Urban Environment Today*, 10 February, p. 9.

Medhat, A. A. and Khalil, E. E. (2006). Thermal comfort meets human acclimatization in Egypt, *Proceeding of Healthy Building*, Vol. 2, p. 25.

Mercer. (2011). *Mercer 2011 Quality of Living Survey Highlights—Defining 'Quality of Living'*, Mercer, Retrieved January 25, 2012, from http://www.mercer.com/articles/quality-of-living-definition-1436405

Milford, E. (2009). Masdar city: A source of inspiration, *Renewable Energy World Magazine*, Vol. 12, No. 2, Retrieved March 15, 2009, from http://www.renewableenergyworld.com/rea/news/article/2009/04/masdar-city-a-source-of-inspiration

Ministry of Economic Development. (2008). *Egypt Achieving The Millennium Development Goals: A Midpoint Assessment*, Ministry of Economic Development, Cairo, Egypt.

Ministry of Housing, Utitilties and Urban Development. (2008). *Project for Giza Northern Sector Development and Reuse of Imbaba Airport Site*, Ministry of Housing, Utilities and Urban Development, Cairo.

Ministry of State for Environmental Affairs and Egyptian Environmental Affairs Agency. (2009). Climate change 2001–2009, Retrieved May 3, 2009, from http://www.eeaa.gov.eg

Moffatt, I. (2000). Ecological footprints and sustainable development, *Ecological Economics*, Vol. 32, No. 3, pp. 359–362.

Moos, M., Whitfield, J., Johnson, L. C., and Andrey, J. (2006). Does design matter? The ecological footprint as a planning tool at the local level, *Journal of Urban Design*, Vol. 11, No. 2, pp. 195–224.

Morris, W. and Kaufman, J. (1998). The new urbanism: An introduction to the movement and its potential impact on travel demand with an outline of its application in Western Australia, *Urban Design International*, Vol. 3, No. 4, pp. 207–221.

Muñoz, S. S. (2007). Do 'green' appliances live up to their billing, *The Wall Street Journal*, p. 360.

Murakami, S. (2008). Promoting eco-model cities to create a low-carbon society, *International Seminar on Promoting the Eco-Model Cities for The Low Carbon Society*, Tokyo, Japan.

NÆSS, P. (1997). *Fysisk Planlegging Og Energibruk* [Physical planning and energy use], Tano Aschehoug, Oslo.

Nakamura, Y. and Fujikawa, A. (2002). Evaluation of thermal comfort and energy conservation of an ecological village office, *ROOMVENT 2002*, pp. 413–416.

Nasar, J. (1997). Neo-traditional development, auto-dependency and sense of community, In M. Amiel and J. Vischer, eds., *Space Design and Management for Place Making*, EDRA 28, EDRA, Edmond, OK, pp. 39–43.

Nation Ranking. (2011). *Quality of Life Index 2011 Rankings*, Nation Ranking: Quantifying the World of Sovereign States, Retrieved January 21, 2012, from https://nationranking.wordpress.com/category/quality-of-life-index/

Naydenov, K., Pitchurov, G., Langkilde, G., and Melikov, A. K. (2002). Performance of displacement ventilation in practice, *ROOMVENT 2002*, pp. 483–486.

Newman, P. (2006). The environmental impact of cities, *Environment and Urbanization*, Vol. 18, No. 2, pp. 275–295.

Newman, P. and Kenworthy, J. (1989a). *Cities and Automobile Dependence: An International Sourcebook*, Gower Technical Press, Brookfield, VT.

Newman, P. and Kenworthy, J. (1989b). Gasoline consumption and cities: A comparison of US cities with a global survey, *Journal of the American Planning Association*, Vol. 55, No. 1, pp. 24–37.

Newman, P. and Kenworthy, J. R. (1999). *Sustainability and Cities: Overcoming Automobile Dependence*, Island Press, Washington, DC.

Ng, J. (2009). *New Energy Star 5.0 Specs for Computers Become Effective Today*. DailyTech. Retrieved July 1, 2009, from http://www.dailytech.com/New+Energy+Star+ 50+Specs+for+Computers+Become+Effective+Today/article15559.htm

Oldfield King Planning Ltd. (1998). *Car Parking and Social Housing: National Planning and Housing Policy*, National Housing Federation, London.

Oldham, L., Shorter, F., and Tekçe, B. (1994). *A Place to Live, Families and Child Health in a Cairo Neighbourhood*, The American University in Cairo Press, Cairo, p. 10.

Olesen, B. W., Koschenz, M., and Johansson, C. (2003). New European standard proposal for design and dimensioning of embedded radiant surface heating and cooling systems, *ASHRAE Transactions*, Vol. 109, pp. 107–120.

Omer, A. M. (2008). Energy, environment and sustainable development, *Renewable and Sustainable Energy Reviews*, Vol. 12, No. 9, pp. 2265–2300.

Orchard, L. (1995). National urban policy in the 1990's, In P. Troy, ed., *Australian Cities: Issues, Strategies and Policies for Urban Australia in The 1990's*, University of Cambridge Press, Cambridge, pp. 65–86.

Organization for Economic Cooperation and Development (OECD). (2011). *Better Life Initiative Executive Summary*, OECD Better Life Initiative, Retrieved January 20, 2012, from http://oecdbetterlifeindex.org/

Our Human Development Initiative. (2011). Global Footprint Network, Retrieved April 30, 2012, from http://www.footprintnetwork.org/en/index.php/GFN/page/fighting_poverty_our_human_development_initiative/

Owens, S. (1992). Energy, environmental sustainability and land use planning, In M. J. Breheny, ed., *Sustainable Development and Urban Form*, Pion Ltd, London, pp. 79–105.

Panão, M. O., Gonçalves, H. J., and Ferrão, P. M. (2008). Optimization of the urban building efficiency potential for mid-latitude climates using a genetic algorithm approach, *Renewable Energy*, Vol. 33, No. 5, pp. 887–896.

Pickup, L. (1984). Women's gender-role and its influence on travel behavior, *Built Environment*, Vol. 10, No. 1, pp. 61–68.

Pollard, T. (2001). Greening the American dream? *Planning*, Vol. 67, No. 10, pp. 10–16.

PowerPulse.net. (1948). New Energy Star Promoting New Specs at APEC and PPDC, Retrieved June 8, 2006, from 80plus.org

PrEN 13363-1, Solar protection devices combined with glazing—Calculation of solar energy and light transmittance—Part 1: Simplified method, Brussels, 2005.

PrEN ISO 10077-2, Thermal performance of windows, doors and shutters—Calculation of thermal transmittance—Part 2: Numerical methods for frames, Brussels, 2005.

Principles of New Urbanism. (n.d.). New Urbanism, Retrieved December 24, 2009, from http://www.newurbanism.org/newurbanism

Radberg, J. (1998). Ebenezer Howard's dream: 100 years after. *Paper presented at the 44th International Federation of Housing and Planning World Congress*, September 13–17, Lisbon, Portugal.

Rapley, M. (2003). *Quality of Life Research: A Critical Introduction*, Sage, London.

Rapport, D. J. (2000). Ecological footprints and ecosystem health: Complementary approaches to a sustainable future, *Ecological Economics*, Vol. 32, No. 3, pp. 367–370.

Rees, W. (1992). Ecological footprints and appropriated carrying capacity, *Environment & Urbanization*, Vol. 4, No. 2, pp. 121–130.

Rees, W. (2000). Eco-footprint analysis: Merits and brickbats, *Ecological Economics*, Vol. 32, No. 3, pp. 371–374.

REN21 Renewable Energy Policy Network for the 21st Century. (2014). *Renewables 2014, Global Status Report*, REN21 Secretariat, UNEP, Paris, France.

Roback, J. (1982). Wages, rents, and the quality of life, *Journal of Political Economy*, Vol. 90, pp. 1257–1278.

Roy, A. (2005). Urban informality: Toward an epistemology of planning, *Journal of The American Planning Association*, Vol. 71, No. 2, pp. 147–158.

Ruta, D. (1998). Patient generated assessment: The next generation, *MAPI Quality of Life Newsletter*, Vol. 20, pp. 461–489.

Ruta, D., Camfield, L., and Martin, F. (2004). Assessing individual quality of life in developing countries: Piloting a global PGI in Ethiopia and Bangladesh. *Quality of Life Research*, Vol. 13, No. 9, p. 1545.

Ruta, D. A., Garratt, A. M., Leng, M., Russell, I. T., and MacDonald, L. M. (1994). A new approach to the measurement of quality of life: The patient generated index (PGI). *Medical Care*, Vol. 32, No. 11, pp. 1109–1126.

Rydin, Y. (1992). Environmental dimensions of residential development and the implications for local planning practice, *Journal of Environmental Planning and Management*, Vol. 35, No. 1, pp. 43–61.

Scully, V. (1994). The architecture of community, In P. Katz, ed., *The New Urbanism: Toward an Architecture of Community*, McGraw-Hill, New York, pp. 221–230.

Self, P. (1997). *Environmentalism and Cities. Newsletter Urban Research Program no. 32*, Research School of Social Sciences, Australian National University, Canberra.

Shannon, G. and Cromley, E. (1985). Settlement and density patterns: Toward the 21st century, In J. Wohlwill and W. Van Vliet, eds., *Habitats for Children: The Impacts of Density*, Lawrence Erlbaum, Hillsdale, NJ, pp. 1–16.

Shea, W. (1976). Introduction: The quest for a high quality of life, In J. King-Farlow and W. Shea, eds., *Values and The Quality of Life*, Science History Publications, New York, pp. 1–5.

Sherlock, H. (1991). *Cities are Good for Us*, Paladin, London.

Shireman, W. (1992). How to use the market to reduce sprawl, congestion and waste in our cities, In B. Walter et al., eds., *Sustainable Cities: Concepts And Strategies for Eco-City Development*, Eco-Home Media, Los Angeles, CA, pp. 1–354.

Simmonds, D. and Coombe, D. (2000). The transport implications of alternative urban forms, In M. Jenks, E. Burton, and K. Williams, eds., *Achieving Sustainable Urban Form*, E & FN Spon, London, pp. 115–123.

Smyth, H. (1996). Running the gauntlet: A compact city within a doughnut of decay, In M. Jenks, E. Burton, and K. Williams, eds., *The Compact City. A Sustainable Urban Form?* E & FN Spon, London, pp. 101–113.

Smyth, J. (1992). The economic power of sustainable development: Building the new American dream, In B. Walter et al., eds., *Sustainable Cities: Concepts And Strategies for Eco-City Development*, Eco-Home Media, Los Angeles, CA, pp. 1–354.

Spangenberg, J. H. and Lorek, S. (2002). Environmentally sustainable household consumption: From aggregate environmental pressures to priority fields of action, *Ecological Economics*, Vol. 43, pp. 127–140.

Stead, D., Williams, J., and Titheridge, H. (2000). Land use, transport and people: Identifying the connections, In M. Jenks, E. Burton, and K. Williams, eds., *Achieving Sustainable Urban Form*, E & FN Spon, London, pp. 174–186.

Stenhouse, D. (1992). Energy conservation benefits of high density mixed-use land development, In B. Walter et al., eds., *Sustainable Cities: Concepts And Strategies for Eco-City Development*, Eco-Home Media, Los Angeles, CA, pp. 1–354.

Stoel, T. B., Jr. (1999). Reining in urban sprawl, *Environment*, Vol. 41, No. 4, pp. 6–11.

Stubbs, M. (2002). Car parking and residential development: Sustainability, design and planning policy, and public perceptions of parking provision, *Journal of Urban Design*, Vol. 7, No. 2, pp. 213–237.

Templet, P. (2000). Externalities, subsidies, and the ecological footprint: An empirical analysis, *Ecological Economics*, Vol. 32, No. 3, pp. 381–384.

The Academy of Urbanism. (2010). *The Frieburg Charter of Sustainable Urbanism: Learning from Place*, The Academy of Urbanism, London.

The Economist Intelligence Unit (EIU). (2007). The Economist Intelligence Unit's quality-of-life index, THE WORLD IN 2005, The Economist, Retrieved January 20, 2012, from http://www.economist.com/media/pdf/QUALITY_OF_LIFE.pdf

The Economist Intelligence Unit. (2009). *European Green City Index: Assessing The Environmental Impact of Europe's Major Cities*, Siemens AG, Munich.

The Economist Intelligence Unit. (2010). *Latin American Green City Index: Assessing The Environmental Performance of Latin America's Major Cities*, Siemens AG, Munich.

The Economist Intelligence Unit. (2011a). *African Green City Index: Assessing The Environmental Performance of Africa's Major Cities*, Siemens AG, Munich.

The Economist Intelligence Unit. (2011b). *Asian Green City Index: Assessing The Environmental Performance of Asia's Major Cities*, Siemens AG, Munich.

The Economist Intelligence Unit. (2011c). *US and Canada Green City Index: Assessing The Environmental Performance of 27 Major US and Canadian Cities*, Siemens AG, Munich.

The Prince's Foundation. (2000). *Sustainable Urban Extensions: Planned Through Design. A Collaborative Approach to Developing Sustainable Town Extensions through Enquiry by Design*, The Prince's Foundation, London, p. 1.

The United Nations Centre for Human Settlements (UNCHS) and the United Nations Environment Programme (UNEP). (1999). *The SCP Source Book Series, Volume 5 Institutionalising the Environmental Planning and Management (EPM) Process*, UNCHS, Nairobi, Kenya.

The United Nations Economic Commission for Africa (UNECA) and North Africa Office. (2003). The fight against desertification and drought in North Africa, *The Eighteenth Meeting of the Intergovernmental Committee of Experts*, United Nations Economic Commission for Africa, Tangiers, Morocco.

Thompson-Fawcett, M. (2000). The contribution of urban villages to sustainable development, In M. Jenks, E. Burton, and K. Williams, eds., *Achieving Sustainable Urban Form*, E & FN Spon, London, pp. 275–285.

Timmer, J. (2008). *EPA Tightens Rules on its Green Power Partners*, Arstechnica.com, Retrieved March 23, 2009, from http://arstechnica.com/old/content/2008/12/epa-tightens-rules-on-its-green-power-partners.ars

Tirado, J. (n.d.). *Smart Growth Advances Nationally*, New Urbanism, Retrieved December 24, 2009, from http://www.newurbanism.org

Titheridge, H., Hall, S., and Banister, D. (2000). Assessing the sustainability of urban development policies, In M. Jenks, E. Burton, and K. Williams, eds., *Achieving Sustainable Urban Form*, E & FN Spon, London, pp. 149–159.

Transit Oriented Development. (n.d.). *Design for a Livable Sustainable Future*, Transit Oriented Development, Retrieved January 29, 2010, from http://www.transitorienteddevelopment.org

Tugend, A. (2008). If your appliances are avocado, they're probably not green, *New York Times*. Retrieved June 29, 2008, from http://www.nytimes.com/2008/05/10/business/yourmoney/10shortcuts.html?scp = 1&sq = appliances%20avocado%20green&st = cse

UN General Assembly. (2014). Report of the Open Working Group of the General Assembly on Sustainable Development Goals, Sixty – eighth session, A/68/970, Retrieved August 15, 2014, from http://undocs.org/A/68/970

United Nations Development Programme (UNDP). (2010). *Human Development Report 2010, 20th Anniversary Edition, The Real Wealth of Nations: Pathways to Human Development*, UNDP, New York.

United Nations Environment Programme (UNEP). *Activities: Urban—Energy for Cities*, Retrieved May 14, 2009, from http://www.unep.or.jp/Ietc/Activities/Urban/energy_city.asp

United Nations Human Settlements Programme (UN-HABITAT). (2006). *The State of The World's Cities Report 2006/2007: 30 Years of Shaping The Habitat Agenda*, Earthscan, London, UK.

United Nations Human Settlements Programme (UN-HABITAT). (2008). *State of the World's Cities 2008/2009: Harmonious Cities*, Earthscan, London, UK.

United Nations Human Settlements Programme (UN-HABITAT). (2009). *Planning Sustainable Cities: Policy Directions, Global Report On Human Settlements*, Abridged Edition, Earthscan, London, UK.

United Nations Human Settlements Programme (UN-HABITAT). (2012a). *Evaluation Report 2/2012 Mid-Term Evaluation of the Cities and Climate Change Initiative*, UN-Habitat, Nairobi.

United Nations Human Settlements Programme (UN-HABITAT). (2012b). *State of the World's Cities Report 2012/2013: Prosperity of Cities*, UN-HABITAT, Nairobi, Kenya.

United Nations Human Settlements Programme (UN-HABITAT). (2013). *State of the World's Cities 2012/2013: Prosperity of Cities*, Routledge, New York.

USA DOE, 2002, Retrieved www.energy.gov

U.S. Environmental Protection Agency. *2006 Annual Report: Energy Star and Other Climate Protection Partnerships*. Retrieved March 1, 2008, from http://www.energystar.gov/ia/news/downloads/annual_report_2006.pdf.

US-Green Building Council, 2012, Retrieved www.usgbc.org

Van den Bergh, J. and Verbruggen, H. (1999). Spatial sustainability, trade and indicators: An evaluation of the ecological footprint, *Ecological Economics*, Vol. 29, pp. 61–72.

Van der Ryn, S. (1986). The suburban context, In S. Van der Ryn and P. Calthorpe, eds., *Sustainable Communities: A New Design Synthesis for Cities, Suburbs and Towns*, Sierra Club Books, San Francisco, CA, pp. 1–238.

Van Dijk, D. and Khalil, E. E. (2009). Energy efficiency in buildings, *ISO Focus*, pp. 16–20.

Van Dijk, D. and Khalil, E. E. (2011). Future cities—Building on energy efficiency, *ISO Focus*, pp. 25–27.

Veenhoven, R. (1996). The study of life satisfaction, In W. Saris, R. Veenhoven, A. C. Scherpenzeel, and B. Bunting, eds., *A Comparative Study of Satisfaction with Life in Europe*, Eötvös University Press, Budapest, pp. 11–48.

Veenhoven, R. (2007). Subjective measures of well-being, In M. McGillivray, ed., *Human Well-being, Concept and Measurement*, Palgrave/McMillan, Houndmills, NH, pp. 214–239.

Vilhelmson, B. (1990). *Va r dagliga rörlighet: om resandes utveckling, fördelning och gränser* [Our Daily Mobility: On the Development, Distribution and Limits of Travelling]. TFB report 1990: 16, The Swedish Transport Board, Stockholm.

Wackernagel, M. (2009). Securing human development in a resource-constrained world, *DAC News*, Retrieved April 30, 2012, from http://www.oecd.org/dataoecd/52/1/43844294.htm#H56

Wackernagel, M. and Rees, W. (1996). *Our Ecological Footprint: Reducing Human Impact on the Earth*, New Society Publishers, Gabriola Island, BC.

Walsh, B. (1997). The right mix for urban living, *Urban Environment Today*, Vol. 14, No. 20, pp. 8–9.

Walter, B., Lois, A. and Crenshaw, R., eds. (1992). *Sustainable Cities: Concepts and Strategies for Eco-City Development*, Eco-Home Media, Los Angeles, CA.

WHOQOL Group. (1998). Development of the World Health Organization WHOQOL-BREF quality of life assessment, *Psychological Medicine*, Vol. 28, pp. 551–558.

Why Obama's Energy Savings Estimate May Be Skewed, New York Times, 2010.

Wilbanks, T. J. and Kates, R. W. (1999). Global change in local places: How scale matters, *Climatic Change* Vol. 43, No. 3, pp. 601–628.

Woodcock, A., Camfield, L., McGregor, J. A., and Martin, F. (2009). Validation of the WeDQoL-goals-Thailand measure: Culture-specific individualized quality of life, *Social Indicators Research*, Vol. 94, No. 1, pp. 135–171

Woodhull, J. (1992). How alternative forms of development can reduce traffic congestion, In B. Walter et al., eds., *Sustainable Cities: Concepts and Strategies for Eco-City Development*, Eco-Home Media, Los Angeles, CA, pp. 1–354.

World Energy Council (WEC). (2013). *World Energy Resources: 2013 Survey*, World Energy Council, London. www.worldenergy.org

www.ecocitizenworldmap.org

www.ecocitystandards.org

www.egypt-urban.net

www.energystar.gov/index.cfm?c=evaluate_performance.pt_neprs_learn

www.energystar.gov/benchmark

www.masdar.ae/en/

Index

Note: Page numbers in *italics* refer to figures and tables.